# 6th International Conference on Architecture, Materials and Construction (ICAMC 2020)



Held online due to COVID-19

IOP Conference Series: Materials Science and Engineering
Volume 1044

Lisbon, Portugal
27 - 29 October 2020

ISBN: 978-1-7138-3332-1
ISSN: 1757-8981

**Printed from e-media with permission by:**

Curran Associates, Inc.
57 Morehouse Lane
Red Hook, NY 12571

**Some format issues inherent in the e-media version may also appear in this print version.**



Institute of Physics
Dirac House, Temple Back
Bristol BS1 6BE UK

Phone: 44 1 17 929 7481
Fax: 44 1 17 920 0979

techtracking@iop.org

**Additional copies of this publication are available from:**

Curran Associates, Inc.
57 Morehouse Lane
Red Hook, NY 12571 USA
Phone: 845-758-0400
Fax: 845-758-2633
Email: curran@proceedings.com
Web: www.proceedings.com

# TABLE OF CONTENTS

IOP Conf. Series: Materials Science and Engineering 1044 (2021) 011001 doi:10.1088/1757-899X/1044/1/011001

# Preface

The University of Lisbon, in partnership with CBEES, organized the 6th International Conference on Architecture, Materials, and Construction (ICAMC 2020) and 1st International Conference on Building Science, Technology and Sustainability (ICBSTS 2020), from October 27 to 29, 2020, with significant scientific community participation.

Due to the COVID-19 outbreak, both conferences, scheduled to be held at the University of Lisbon, Portugal, took place on the Zoom platform, guaranteeing all participants' health and safety. The organizers resort to high-speed internet, professional video-recorder, and technical audio support to secure all sessions the best live experience, promoting an interactive and enthusiastic discussion between different background participants during Q&A periods.

The Conference complied with the scheduling requirements by presenting 20 specialized reports during three sessions and four plenary speeches. Aside from the selected University of Lisbon's scientists, the scientific forums gathered worldwide interest, like the U.S., U.K., Greece, Poland, Peru, among others who have contributed significantly to overall success. Plenary speakers had 40 minutes to present, including 5 minutes for the Q&A, and the sessions had 15 minutes for each presentation, including 2 minutes for the Q&A. The University of Lisbon promoted its master students' attendance and participation, enriching in-house curriculums by enabling interaction with the Conferences' participants.

The main goal of both events, ICAMC & ICBSTS, is to bring together academic scientists, engineers, and industry researchers to exchange their experiences and research results. Also, discuss the challenges, solutions adopted, and practical applications of the exhibited technology and knowledge. All papers in this proceeding were subject to peer-review by conference committee members and international reviewers. Conferences' relevance and quality drive the selection of the documents included in the proceedings.

Finally, I applaud the authors on the submitted papers' novelty and reviewers' dedicated time to improve the final result. My sincere congratulation to the staff for making everything seem incredibly smooth and easy. Greetings to the proceeding team for the dedication, support, and countless time to bring these scratches into this volume.

I sincerely hope to meet you in person in the upcoming editions.

Prof. Nuno Dinis Cortiços
Universidade de Lisboa, Portugal

**Conference Chair**
Prof. Nuno Dinis Cortiços, Universidade de Lisboa, Portugal

**Program Co-Chairs**
Prof. Eric J. Strauss, Michigan State University, USAProf. Marco Casini, SAPIENZA University of Rome, Italy

**Publicity Chair**

IOP Conf. Series: Materials Science and Engineering 1044 (2021) 011001 doi:10.1088/1757-899X/1044/1/011001

Assoc. Prof. Mohammad Arif Kamal, Aligarh Muslim University, India

**International Technical Committee**

Ayse Ozcan,Michigan State University, USARaffaella Maddaluno , Lisbon School of Design - Ciaud , PortugalQi Zhou, Southeast University, ChinaAntonio Rodero,University of Cordoba, SpainAna Almerich-Chulia, Universitat Politecnica de Valencia, SpainKatarzyna Gładyszewska-Fiedoruk, Bialystok University of Technology, Poland Dorota Anna Krawczyk,Bialystok University of Technology , Poland Fouad K. Kaddah, Université Saint-Joseph de Beyrouth, LebanonWassim E. Raphael, Université Saint-Joseph de Beyrouth, LebanonElise ZGHEIB, Lebanese University, Hadath, Lebanon Juan Fernando Valencia,Saint Thomas University, ColombiaKhaled M. Dewidar, British University, Cairo, EgyptSalim Ferwati, Qatar University, QatarHUYNH, TRONG-PHUOC, Can Tho University, VietnamWegdan Wagdy El-Nadoury, Pharos University, Alexandria, EgyptUzma Khan,Manipal School of Architecture and Planning, MAHE, IndiaShady Salem,The British University in EgyptChung-Suk Cho, Khalifa University,Abu Dhabi, UAESyed Ishtiaq Ahmad, Bangladesh University of Engineering and Technology , Bangladesh Owus Ibearugbulem,Civil Engineering Department Federal Polytechnic Nekede Owerri Imo State NigeriaSamuel I. Egwunatum,Owerri Imo State, Nigeria

IOP Conf. Series: Materials Science and Engineering 1044 (2021) 011002 doi:10.1088/1757-899X/1044/1/011002

# Peer review declaration

All papers published in this volume of *IOP Conference Series: Materials Science and Engineering* have been peer reviewed through processes administered by the Editors. Reviews were conducted by expert referees to the professional and scientific standards expected of a proceedings journal published by IOP Publishing.

- **Type of peer review:** Double-blind
- **Conference submission management system:** iConf Conference System
- **Number of submissions received: 36** (26 Full paper + 10 Abstract)
- **Number of submissions sent for review:** 26
- **Number of submissions accepted:** 21
- **Acceptance Rate (Number of Submissions Accepted / Number of Submissions Received X 100):** 58%
- **Average number of reviews per paper:** 2
- **Total number of reviewers involved:** 21
- **Any additional info on review process:**
  All received submissions were subjected to double-blind peer reviews, who are expert or have been experiencing in the related field for years. The accepted papers must be revised according to the referees' comments and suggestions, before inclusion in the conference proceedings. Manuscripts submitted should be original & unpublished and written in accordance with the standard IOP paper template.
- **Contact person for queries:** Ms. Violet Xie (icamc@cbees.net)

IOP Conf. Series: Materials Science and Engineering 1044 (2021) 012001 doi:10.1088/1757-899X/1044/1/012001

# Thermal Insulation Performance Optimization of Hollow Bricks Made up of 3D Printable Rubber-Cement Mortars: Material Properties and FEM-based Modelling

**M Sambucci**[1,2,*] **and M Valente**[1,2]

[1] Department of Chemical and Material Engineering, Sapienza University of Rome, 00184 Rome, Italy
[2] INSTM Reference Laboratory for Engineering of Surface Treatments, Department of Chemical and Material Engineering, Sapienza University of Rome, 00184 Rome, Italy

[*]matteo.sambucci@uniroma1.it

**Abstract.** Sustainable construction is attracting more attention lately. It involves the design of eco-friendly building structures, the reduction of energy consumption and waste, the use of renewable and recyclable resources, and the enhancement of low-impact manufacturing methods. This paper addresses some of these sustainability questions, in the context of the use of tire recycled rubber particles as aggregates of cement mortars suitable for Additive Manufacturing (AM) processes. Specifically, the effect of rubber aggregates on physical and thermal properties was investigated, to evaluate the heat-insulating performance of the compounds. The lightweight and non-polar nature of rubber improve the thermal insulation and physical-structural properties of the material, in terms of thermal conductivity, unit weight, and porosity respectively. However, these effects are closely related to the particle size and their adhesion with the cement matrix. In the second part of the manuscript, applicability study of rubber-cement compounds based on the design and finite element method (FEM)-based thermal analysis of innovative hollow bricks is presented. Fractal cavities were investigated as a functional inner architecture to improve the thermal behavior of the component. FEM results show an increase of more than 30% in thermal resistance ($R_T$) for fractal-based brick compared to conventional designs, demonstrating that the holes' geometric irregularity is a key feature in the thermal flow attenuation.

## 1 Introduction

Brick is one of the main building materials of masonry structures. Since the ancient Romans period, no other material has managed to replace it, due to its durability, insulating power, cheapness, and mechanical strength. Nowadays, the rising demand for building materials, the increase in demolition wastes, and the impact of brick manufacturing processes on the environment and human health have encouraged the development of sustainable architecture and construction [1]. These approaches are focused on the design of structures and architectural units with low environmental impact, minimizing the use of natural resources, and enhancing energy efficiency. In this framework, the use of industrial wastes as aggregates of cement mortars can represent a noteworthy strategy in terms of sustainability and functional properties. In the literature, several examples on the reuse of recycled fillers in cementitious compounds can be found, including PET particles [2], wood chips [3], glass powder [4], and waste tire rubber aggregates (TRA) [5]. The reuse of shredded TRA as building materials constituents is a widely studied solution by researchers to counteract the environmental pollution, accumulation of end-of-life tires, and to design lightweight materials with improved ductility, durability, and thermo-acoustic insulation properties [6,7].

This work focuses on the experimental studies related to the development of technologically advanced cementitious mortars suitable for Additive Manufacturing (AM) process modified with polymer particles deriving from waste tires. The possibility of using the rubberized compounds in 3D printing technology represents the novel aspect of this research. As well investigated, AM in the construction sector implies interesting benefits [8]: a) building materials savings and demolition wastes reduction; b) lowering in injury rates; c) high architectural flexibility. According to the requirements mentioned above, an attractive perspective opens up in the design of more eco-friendly and engineering-functional building components.

To investigate the peculiarities of these mixtures as insulation solutions in the construction field, the effect of tire rubber on thermal performance was analyzed by physical and thermal experimental testing. First, three different printable rubber-cement compounds (100 % sand-rubber replacement level) varying the size of TRA (0-1 mm rubber powder and 2-4 mm rubber granules) were studied. Control mortar (CTR) containing 100 % sand was also prepared to compare its properties with those of rubber-modified samples. Finally, the authors present a Finite Element Method (FEM)-based thermal analysis on the technological applicability of these materials for the manufacturing of multi-perforated bricks. Specifically, different hole designs were examined and then evaluated the heat insulation performance of the bricks in terms of thermal resistance ($R_T$). Innovative inner architectures based on hexagonal and fractal cavities were investigated and compared with conventional configurations. As suggested by a previous analytical study [6], hexagonal cells improve the mechanical deformability behavior and the sound insulation performance of the brick, maintaining satisfactory structural properties. On the other hand, fractal cavities could represent an ideal candidate for the thermal optimization of bricks, due to the marked geometry complexity. As confirmed by Bustamante et al. [9], the increase in the shape complexity of the holes improves the thermal behavior of hollow blocks, because the thermal trajectory is maximized and therefore the convection heat transfer is reduced. The research's purpose is to obtain a "bifunctional" building element, having optimized insulating properties deriving both from the material composition and its topological features.

## 2. Materials and Methods

### *2.1. Materials*

3D printable rubber-cement mixtures were obtained starting from a control mortar (CTR) consisting of Type I Portland cement, limestone sand of 0.4 mm nominal maximum size, and a specific blend of chemical additives (Silica fume-based thixotropic additive, Polycarboxylate ether-based superplasticizer, Aliphatic-based water reducing agent, and Calcium oxide-based expansive agents). Two type of TRA from the mechanically grinding process of waste tires, sourced by European Tyre Recycling Association (ETRA), were used as a total replacement of sand by volume: 0-1 mm rubber powder (RP) and 2-4 mm rubber granules (RG). The average density of rubber aggregates, evaluated by Micromeritics AccuPyc 1330 He-pycnometer, is 1202 kg/m$^3$. By varying the replacement ratios of RP and RG, three rubberized mortars were developed and tested.

As shown in table 1, water dosage is the only parameter that is altered in the mixes production. The proper mixes rheology was selected following specific printability requirements [8]: extrusion process without interruptions (*extrudability*), stacking integrity between printed layers (*buildability*), and strong layer-by-layer binding (*inter-layers adhesion*). These 3D printing indicators were evaluated by printability tests described in the next section.

Table 1 presents the mix compositions for CTR mix and rubber-cement compounds per m$^3$.

**Table 1.** Mix proportions for CTR and rubberized mortars

| Mix design | *CTR* | *RP100* | *RP50RG50* | *RP25RG75* |
|---|---|---|---|---|
| RP content [% v/v] | 0 | 100 | 50 | 25 |
| RG content [% v/v] | 0 | 0 | 50 | 75 |
| Cement [kg] | 800 | 800 | 800 | 800 |
| Water [kg] | 300 | 260 | 250 | 230 |
| Sand [kg] | 1100 | 0 | 0 | 0 |
| RP [kg] | 0 | 300 | 150 | 75 |
| RG [kg] | 0 | 0 | 160 | 240 |
| Additives [kg] | 152 | 152 | 152 | 152 |

### *2.2. Printability test and specimens manufacturing*

Printability tests were conducted to evaluate the print quality of the mixtures investigated in this research. For this purpose, a COMAU 3-axis robotic arm-based printing apparatus equipped with a circular nozzle (Ø = 10 mm) was adopted. Dry components were mixed in an Aluminum tank for 45 s. Water was then added to the dry compound and mixed for 12 min with an electric mixer drill. After this, the fresh material is deposited in the delivery system for printing. The extrusion speed and the deposition pressure were fixed at 33 mm/s and 4 bar, respectively. Six-layers slabs (230 mm x 160 mm x 55 mm) were printed using each mixture to test the extrudability, buildability, and inter-layer adhesion requirements. Specimens used for the thermo-physical characterization of the materials were extracted from these printed structures by sawing with a diamond blade. Specifically, four blocks with dimensions 48 mm x 42 mm x 22 mm for each mix were used for laboratory testing. After extraction, a lapping treatment was performed on the specimens, using a Buehler 180 grit carbide abrasive disk, to obtain flat and regular faces. The stages related to the specimens manufacturing are summarized in figure 1.

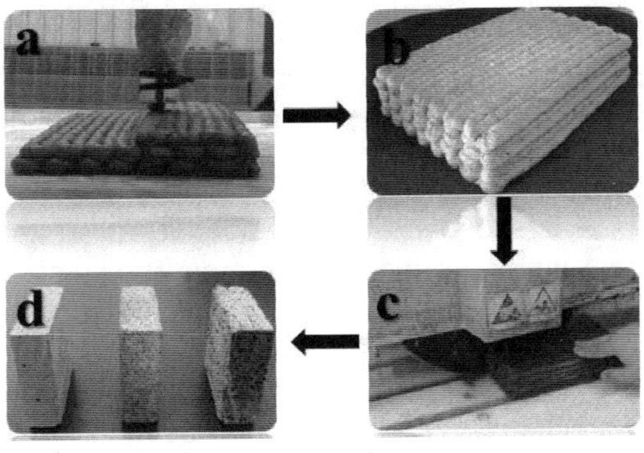


**Figure 1.** Specimens manufacturing process: printability test (a), slabs manufacturing (b), slabs cutting (c), and post-lapping samples (d).

### *2.3. Laboratory tests*

Density, porosity, and heat transfer properties are key physical quantities to describe the thermal behaviour of the materials. In this regard, a series of laboratory tests were conducted to evaluate the effect of TRA on the thermo-physical properties of the cement-based compounds. Furthermore, the experimental results were used as input data for the FEM-based thermal analysis on the hollow bricks optimization described below.

The bulk density ($\rho_b$) of the samples was determined by the mass and volumes of the oven-dried blocks (110 °C for 48 h) in accordance with BS 1881-114 standard [10].

Vacuum saturation technique, as described in ASTM C1202 [11], was conducted to evaluate the permeable porosity ($\varphi$). The sample weight was measured after three physical treatments: 1) oven treatment (110 °C for 2 h); 2) vacuum saturation treatment (0.3 bar for 3 h); 3) water saturation treatment (underwater soaking for 24 h). $\varphi$ was computed considering the weight gain due to water absorption. In support of $\varphi$ measurements, the microstructure and pore distribution of the samples were inspected by microscopy analysis using a Leica MS5 stereomicroscope.

Thermal transport properties of CTR and rubberized specimens, including thermal conductivity (k) and heat capacity ($C_p$), were measured using a C-Therm TCi Thermal Conductivity analyzer (figure 2). According to Modified Transient Plane Source (MTPS) method (ASTM D7984) [12], a one-sided thermal sensor applied a low energy current pulse for 0.8 s to the specimen, so that the k measurement was not affected by thermal convection. Some of the generated heat is absorbed by the specimens and the remaining part is used to increase the temperature at the sensor-sample interface. The temperature variation implies a resistance change of the sensor, resulting in a voltage drop. Voltage data is used to evaluate the thermal transport properties of the investigated material. k and $C_p$ results were obtained from 5 averaged measurement in each mix.

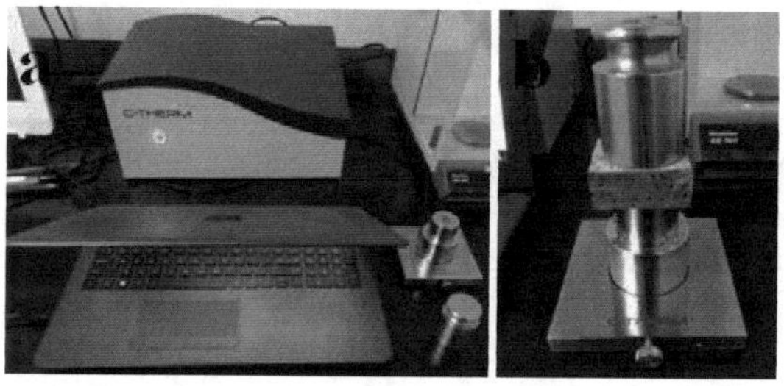

**Figure 2.** Thermal characterization of rubber-cement samples: C-Therm TCi analyzer (a), and test configuration (b).

### 2.4. *FEM-based thermal analysis on rubber-cement hollow bricks*

Hollow bricks are energy-efficient building components because of their combined lightweight, deformability, thermal capacitance, sound insulation, and durability [6]. To reduce the energy consumption and improve the safety and comfort of the structures, a number of studies in the building and architectural fields have focused on increasing the performance of construction components and units. Thanks to the development of FEM techniques and advanced manufacturing methods (such as AM), topological optimization can be considered a valuable approach to model and prototype building elements, having a functional design. As suggested by some researches [9,13], by varying the cavity shapes of a brick its thermal inertia can be modulated. However, traditional manufacturing techniques do not ensure great design freedom and therefore the applications in this regard are very limited. On the other hand, combining AM with cementitious materials is an attractive approach to broaden research and studies on shapes and geometries with improved engineering functionality.

In this manuscript's section, the authors present an applicability study of rubber-cement mixtures for the development of hollow bricks with innovative inner designs. Specifically, the thermal insulation behaviour of bricks based on hexagonal and fractal geometry cavities was simulated and compared with common brick designs.

*2.4.1. Bricks geometry.* The FEM-based investigation of $R_T$ of hollow bricks was performed on 6 three-dimensional models. The prototypes had the same external dimension (250 mm x 150 mm x 55 mm) and hole concentration ratio (~ 40 %), but different inner design. The top views of the brick models are shown in figure 3.

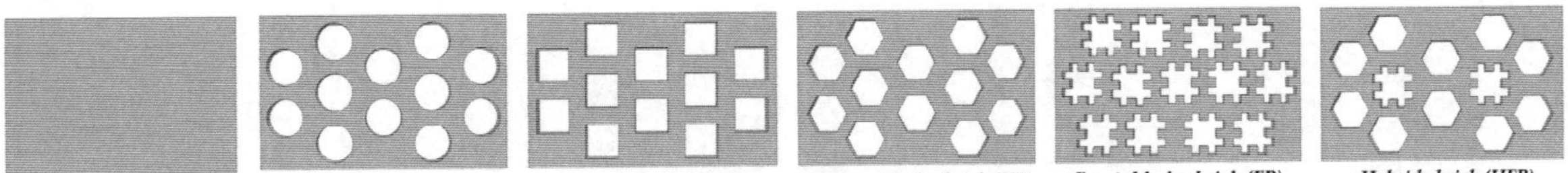


**Figure 3**. Design of the multi-holed bricks

UB identified the unperforated block. CB model is composed of circular holes with 13 mm radius. In SB block, the side of the squares is 33 mm. The hexagons side in HB model is 20.5 mm. The edge of the fractal structures in FB brick is 8 mm. Finally, HFB combined block consisting of an array of 10 hexagonal holes containing two fractal cavities, was tested.

*2.4.2. FEM model for heat transfer in hollow bricks.* FEM-based analysis on the thermal performance of the brick prototypes was performed using the Heat Transfer in Solids Module of COMSOL Multiphysics 5.4 software. $R_T$ was determined through steady-state thermal analysis, as suggested by ISO 9869 standard. The procedure involves the application of temperature gradient ($\Delta T$) across two brick faces perpendicular to a spatial direction, keeping the other boundaries in thermal insulation condition (no heat flux across the surface). These temperatures comply the average summer and winter climatic conditions in Italy and accurately reproduce the environmental circumstances to which the building elements are subjected. In summer conditions, the outdoor ($T_{out}$) and the indoor ($T_{in}$) temperatures were set equal to 40°C and 26°C, respectively. In winter conditions, $T_{out}$ and $T_{in}$ were fixed to 0°C and 20°C, respectively. Besides, the same values of heat flux coefficients were used in both summer and winter states, due to the negligible divergence (~ 3%) in the two investigated climatic circumstances. Outdoor heat flux coefficient ($h_{out}$) and indoor heat flux coefficient ($h_{in}$) were set equal to 24 W/m$^2$K and 7.69 W/m$^2$K, defined by ISO 6946 method. To improve the FEM study, additional thermal ambient properties were selected: average wind velocity ($v_w$), clear sky normal irradiance ($I_{sn}$), and clear sky diffuse irradiance ($I_{sd}$). $v_w$ was taken to be 3.6 m/s [14]. Summer and winter $I_{sn}$ were set as 236 W/m$^2$ and 64 W/m$^2$, respectively. Summer and winter $I_{sd}$ were fixed equal to 93 W/m$^2$ and 28 W/m$^2$, respectively. Irradiance data were deduced by [15], averaging the experimental values, measured in Italy, in summer (May, June, July, and August) and winter (November, December, January, and February) periods. FEM simulation is based on the application of a convective heat flux density on the outdoor ($q_{in}$) and indoor ($q_{out}$) faces of the brick (W/m$^2$). Indoor and outdoor heat flows extracted by convection follows Newton's law of heat (1,2):

$$q_{in} = h_{in} \times \left(T_{in} - T_{if}\right) \tag{1}$$

$$q_{out} = h_{out} \times (T_{out} - T_{of}) \tag{2}$$

where $T_{if}$ and $T_{of}$ are the indoor and outdoor boundary surface temperatures (K), respectively. In steady-state condition, the heat exchange and transmission processes are governed by the thermal Ohm's law (3), which state that the thermal power Q (W) is directly proportional to the temperature gradient ($T_{if} - T_{of}$) causing the thermal exchange:

$$Q = \frac{A}{R_T} \times (T_{if} - T_{of}) \tag{3}$$

where A is the cross-sectional area perpendicular to the path of heat flow (m$^2$). $R_T$ (m$^2$K/W) is strongly related to the thermal inertia of the bricks, which depends on the specific heat capacity, the thermal conductivity, and density of the material. The configuration for thermal simulation of hollows bricks performance is presented in figure 4.

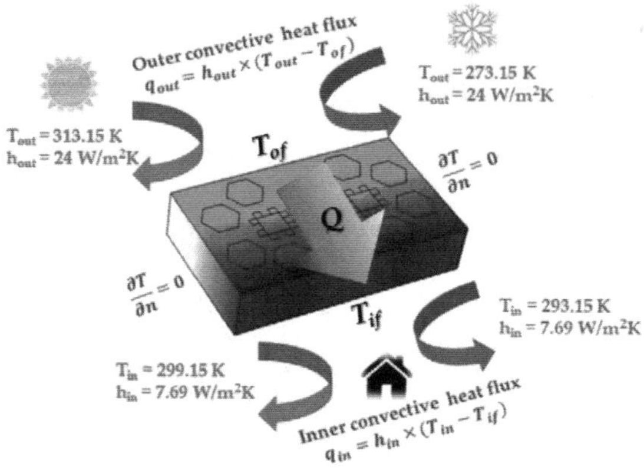


**Figure 4**. FEM-based thermal analysis: simulation configurations

By using Q, $T_{if}$, and $T_{of}$ calculated through the software, it is possible to estimate the $R_T$ of the bricks.

*2.4.3. Materials and meshing*. Hollow brick models are composed of two domains: solid material and cavity. For the solid domain, the thermo-physical properties ($\rho_b$, $\varphi$, k, and $C_p$) of the cement mixtures evaluated in the experimental tests were set. The holes were simulated as air domains by setting the fluid properties from the COMSOL materials library. A fine free-tetrahedral mesh (maximum element size of 20 mm) was used to discretize the entire model

## 3. Results and Discussions

### *3.1 Thermo-physical properties of rubber-cement mortars*

Generally, density and porosity of cementitious materials depend on the aggregate unit weight of aggregates, size of dry constituents, entrapped air, water-cement ratio, and aggregate-cement paste adhesion. As shown in table 2, replacing sand with tire rubber-based lightweight aggregates implies a reduction in $\rho_b$ and $\varphi$ of the cement mortar.

**Table 2.** Density and porosity of mortars with aggregates replaced by rubber particles

| Sample | $\rho_b$ (kg/m³) | $\rho_b$ reduction (%) | $\varphi$(%) | $\varphi$ reduction (%) |
|---|---|---|---|---|
| ***CTR*** | 1927 | / | 33.1 | / |
| ***RP100*** | 1340 | - 30.5 | 22.1 | - 33.2 |
| ***RP50RG50*** | 1624 | - 15.7 | 26.1 | - 21.1 |
| ***RP25RG75*** | 1468 | - 23.8 | 25.7 | - 22.4 |

Changing of physical properties is also due to the rubber particles' tendency to entrap air in their rough surface, as a consequence of their hydrophobic features. However, the non-polar nature of TRA affects the rheological properties of the mixtures. As reported in Table 1, the rubberized compounds consist of lower water dosages than the CTR mix, resulting in lower capillary porosity and air bubbles fraction [16]. This is verified by microstructural analysis, using optical stereomicroscope, reported in figure 5.

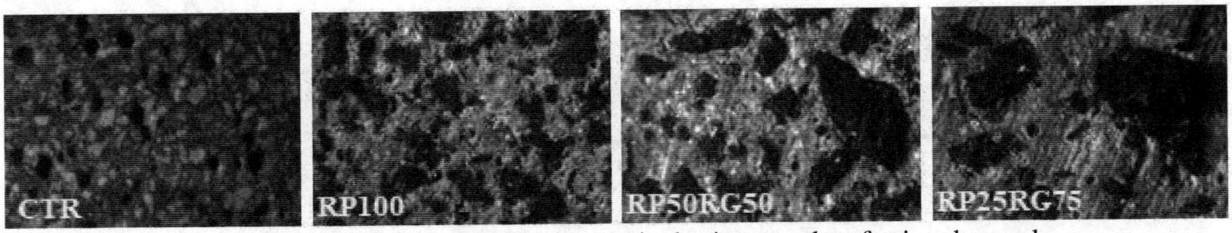


**Figure 5**. 16x magnification optical micrographs of printed samples

By comparing the rubberized mixes behavior, the reduction rates follow a very random trend. Indeed, several competitive factors affect the physical properties of the compounds: a) the effective amount of rubber incorporated into the material; b) the different air adsorption tendency of TRA; c) the rubber-cement adhesion properties. Due to a high specific surface, RP promotes better cohesion with the cement paste but exhibits a greater aptitude for incorporating air. RG allow keeping the water-cement ratio at a low level, hindering the formation of capillary porosity. However, they promote the formation of interface voids due to poor bonding with the matrix [6,7].

Thermal characterization results, given in figure 6, show that the addition of TRA considerably reduce the k-value of the rubber-based mixtures. This decrease is mainly attributable to the higher thermal insulation behavior of the rubber particles than mineral aggregates. Besides, the ability of TRA to entrap air and the material's interfacial porosity are beneficial to obtain low heat conductivity. Due to the relevant thermal inertia of the air, voids oppose the thermal transfer through the matrix, reducing the thermal bridge effects [17]. According to the technical indications given by Newman and Owens [18], rubberized mortars can be classified as lightweight insulating cement materials: oven-dry density range of approximately 300 to a maximum of 2000 kg/m$^3$ and k-value of 0.2 to 1.0 W/m K.

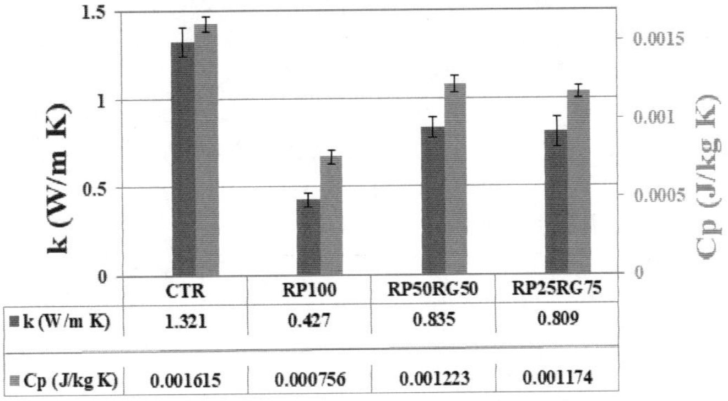


**Figure 6**. k and $C_p$ experimental results

Finally, it is possible to observe that k and $C_p$ varied in the same direction. This is agreeing well with the direct proportionality between the two thermal quantities, well highlighted in the following equation (4):

$$C_p = \frac{k}{\rho_b \times \alpha} \tag{4}$$

where $\alpha$ is the material diffusivity (m$^2$/s).

*3.2 Thermal performance of rubber-cement brick model: FEM analysis results*

$R_T$ is one of the key parameters to analyze the response of a building structure to specific environmental conditions and therefore evaluate its energy performance [14]. The results obtained using FEM are processed in order to obtain the overall $R_T$ for each type of brick design and printable cement mortar (figure 7). The calculation of this parameter was performed in the summer and winter conditions discussed above. The analytical data, determined in the two environmental states, were averaged to provide univocal $R_T$-value.

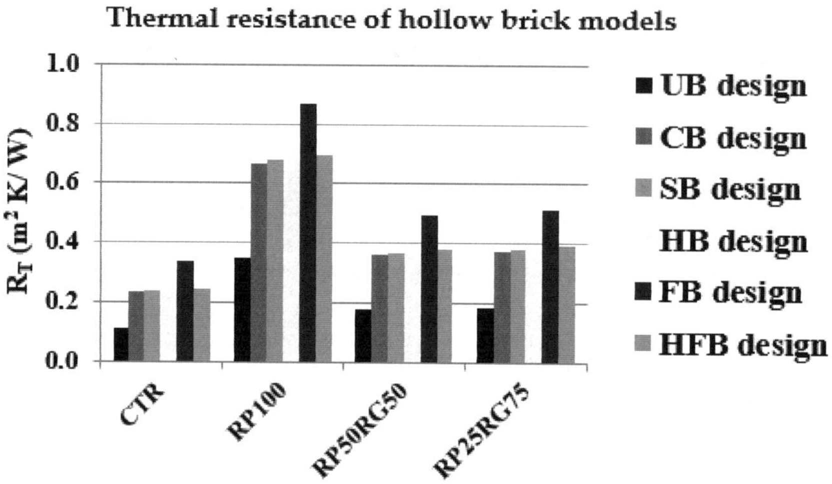


**Figure 7**. FEM results for the rubber-cement bricks of the overall $R_T$

Three relevant results can be deduced from the numerical thermal analysis:

1. Regardless of brick design, $R_T$ is closely related to the heat-insulating properties of the material. The highest $R_T$-values are observable for RP100 mix, where lower k-values were experimentally found compared to the other printable mortars.
2. As expected, inner holes significantly improve the thermal insulation performance of the component. By considering perforated models, it is possible to observe $R_T$-values up to three times higher than UB design, depending on the hollow architecture and cement material. The air cavities act as high-thermal inertia sites, opposing the thermal flow through the material, and enhancing its $R_T$.
3. In each cement compound, $R_T$ performance of FB design is the best. As reported in [6,9], the high geometric complexity of the inner cavities maximizes the thermal path of heat flow, resulting in greater thermal attenuation than conventional architectures. *FB* brick implies a 37%, 34% and 35% average increase in $R_T$ over the CB, SB, and HB models, respectively. However, the fractal cavities morphology may not be optimal in terms of structural performance, since the small curvature radius zones represent preferential triggering points for structural cracks when the brick is subjected to mechanical load. To compensate this potential structural weakness, the authors propose HFB geometry that combines the mechanical-acoustic efficiency of hexagonal holes with the thermal performance of the fractal ones. As know, hexagonal architectures are efficient structures for bending stiffness, due to their aptitude to dissipate residual mechanical stress under load [6]. As shown in figure 7, better thermal insulation response is observed for the hybrid design than conventional geometries, resulting in a potential good matching between structural and thermo-acoustic properties.

## 4. Conclusions and future work

In the first part of this research work, the thermo-physical performances of 3D printable tire rubber-cement mortars were studied. Rubberized mixtures exhibit superior lightweight, microstructural, and thermal insulation properties than plain cement sample as measured by the decrease in density ($\rho_b$), porosity ($\varphi$), and thermal conductivity (k) values, respectively. The percentage-wise improvements on insulation, porosity degree, and unit weight are strictly dependent on the non-polar nature and size of the polymer aggregates that replace the sand. Finally, a numerical thermal analysis technique (FEM) was employed to perform an optimization design study on rubber-cement hollow brick prototypes. To improve the thermal resistance ($R_T$) of building components innovative inner designs based on fractal and hexagonal cavities were tested. Fractal holes-based brick (FB) shows optimum thermal behavior compared to conventional architectures. Due to its geometrical complexity, fractal cavities maximize the thermal trajectory of the heat waves causing relevant attenuation. Hybrid design (HFB) shows intermediate thermal insulation performance between FB and conventional designs but could be more efficient in terms of mechanical properties. Future researches will be focused on evaluating the feasibility in bricks AM, validating the numerical results with laboratory tests, and conducting further studies (FEM and experimental) on the mechanical and acoustic performance of the models.

**Acknowledgments**
The authors express their sincere gratitude to prof. Valeria Corinaldesi and Dr. Glauco Merlonetti (Polytechnic University of Marche) for the support in the additive manufacturing processes.

**References**
[1] Petrillo A, Cioffi R, Ferone C, Colangelo F and Borrelli C, Eco-sustainable geopolymer concrete blocks production process 2015 *Agriculture and agricultural science procedia* **8**(8) 408-18
[2] Saikia N and de Brito J., Mechanical properties and abrasion behaviour of concrete containing shredded PET bottle waste as a partial substitution of natural aggregate 2014 *Construction and building materials* **52** 236-44
[3] Corinaldesi V, Mazzoli A and Siddique R, Characterization of lightweight mortars containing wood processing by-products waste 2016 *Construction and Building Materials* **123**, 281-89
[4] Du H and Tan K H, Properties of high volume glass powder concrete 2017 *Cement and Concrete Composites* **75** 22-29
[5] Siddika A, Al Mamun M A, Alyousef R, Amran Y M, Aslani F and Alabduljabbar H, Properties and utilizations of waste tire rubber in concrete: A review 2019 *Construction and Building Materials* **224** 711-31
[6] Valente M Sambucci M, Sibai A and Musacchi E, Multi-Physics Analysis for Rubber-Cement Applications in Building and Architectural Fields: a Preliminary Analysis 2020 *Sustainability*, **12**(15) 5993
[7] Sambucci M, Valente M, Sibai A, Marini D, Quitadamo A and Musacchi E 2020 Rubber-Cement Composites for Additive Manufacturing: Physical, Mechanical and Thermo-Acoustic Characterization 2020 In *Bos F., Lucas S., Wolfs R., Salet T. (eds) Second RILEM International Conference on Concrete and Digital Fabrication. DC 2020.* RILEM Bookseries, vol 28. Springer, Cham
[8] Valente M Sibai A and Sambucci M, Extrusion-Based additive manufacturing of concrete products: Revolutionizing and remodeling the construction industry 2019 *Journal of Composites Science* **3**(3) 88
[9] Bustamante W, Bobadilla A, Navarrete B, Vidal S and Saelzer G. Thermal improvement of perforated ceramic bricks 2009 *Revista de la Construcción* **8** 24-35
[10] British Standards Institution. BS 1881-114: Testing concrete. Method of determination of density of hardened concrete-1983
[11] American Society for Testing and Materials. ASTM C 1202: Standard Test Method for Electrical Indication of Concrete's Ability to Resist Chloride Ion Penetration-2002
[12] American Society for Testing and Materials. ASTM D 7984-16: Standard Test Method for Measurement of Thermal Effusivity of Fabrics Using a Modified Transient Plane Source (MTPS) Instrument-2016
[13] Arendt K, Krzaczek M, and Florczuk, J, Numerical analysis by FEM and analytical study of the dynamic thermal behavior of hollow bricks with different cavity concentration 2011 *International Journal of Thermal Sciences* **50**(8) 1543-53
[14] Guattari C, Evangelisti L, Gori P and Asdrubali F, Influence of internal heat sources on thermal resistance evaluation through the heat flow meter method 2017 *Energy and Buildings* **135** 187-200

IOP Conf. Series: Materials Science and Engineering 1044 (2021) 012001 doi:10.1088/1757-899X/1044/1/012001

[15] Pasquale V, Solar irradiance in NW Italy 1987 *Theoretical and applied climatology* **38**(2) 85-92
[16] Ghizdăveț Z, Ștefan B M, Nastac D, Vasile O and Bratu M, Sound absorbing materials made by embedding crumb rubber waste in a concrete matrix 2016 *Construction and Building Materials* **124** 755-63
[17] Abu-Jdayil B, Mourad A H and Hussain A, Thermal and physical characteristics of polyester–scrap tire composites 2006 *Construction and Building Materials* **105** 472-79
[18] Owens P, Newman J and Choo B S 2003 *Advanced Concrete Technology Processes*, 1st ed. Oxford, U.K.: Elsevier Ltd., 2003, chapter 2 pp 72-74.

IOP Conf. Series: Materials Science and Engineering 1044 (2021) 012002 doi:10.1088/1757-899X/1044/1/012002

# Evaluation of Material Modification using PET in 3D Concrete Printing Technology

**S Skibicki[1,*], M Pułtorak[1] and M Kaszyńska[1]**

[1]West Pomeranian University of Technology, al. Piastów 17, 70-310 Szczecin, Poland

szymon.skibicki@zut.edu.pl

**Abstract.** Use of recycled materials and incorporation of sustainable development ideas are one of the most popular trends in civil engineering. The 3D concrete printing technology which has been developing rapidly in recent years can benefit from use of both of them. The study evaluates the possibility of using PET granules in mixes suitable for 3D printing. The study compares the properties of two mixes, a reference one (well-studied mix which is suitable for 3D printing) and mix modified by replacing the aggregate with PET (30% of volume). The concretes were tested to evaluate their mechanical properties. The samples for flexural and compressive strength determination were prepared as traditional prisms (40x40x160 mm) as well as printed out with similar dimensions. Additionally, a hollow cylinder structure was printed out and tested for its strength after 24 h. Obtained results has shown that replacement of the aggregate with PET granules did not inhibit the suitability of mix for 3D concrete printing. Use of PET granules decreases the mechanical properties of printed specimens by few to several percent. Presented results can be considered as an initial study for introduction of recommendations for use of PET in printed concretes.

## 1. Introduction

Increasing costs of labour force, the need to speed-up the process of construction and sometimes harmful work conditions proved the need for automation in construction, particularly 3D printing of cement composites. Only in the last decade the number of studies in this area grown exponentially. The basic problem of this technology is the determination of proper composition of the concrete mix. Various research teams have already developed guidelines for concrete mix design [1–8]. Another important issue is the environmental impact of the technology. The idea of sustainable development recommends solutions that reduce greenhouse gas emissions and use of recycled materials [9]. The 3D printing technology can reduce the construction time [10–12] and material consumption [13–15] while also reducing the total amount of waste [16–18]. The above mentioned advantages result from the printing technology itself and not from the mix composition. However, the most important changes beneficial for the environment are the modifications to the mix itself. The modifications to the composition can be done either to the binder [18–21] or aggregate [22]. The use of recycled plastic waste as a partial replacement of the binder or aggregate seems to be a good solution. Today, as much as 348 million tons of plastic are produced annually for candles (data from 2017 according to [23]), of which less than 10% is recycled (data from 2015 according to [23]). The use of plastic waste (e.g. PET granules) in concrete was already studied [24, 25]. Theoretically, it should be possible to use it in printed concretes. As part of the study, a reference mix and mix modified with PET granules were prepared. Both mixtures were tested for their suitability for 3D printing. Presented research is an introduction to the development of recommendations on the use of PET waste in printed concretes.

(a) (b)

**Figure 1.** Examples of specimens prepared by 3D concrete printing technology: a) specimen during printing process; b) printed Specimen after compression test.

## 2. Materials and experimental procedure

### *2.1. Mix composition*

Study was based on two mixes. The first mix B800/PET0 was a reference mix that is suitable for 3D printing and was already tested in other studies [7, 22]. The second (alternative) mix B800/PET30 was modified by replacing 30% of fine aggregate volume with PET granules. In both mixes the water/binder ratio was assumed as constant value of 0.3. The binder in each mix consist of 70% of cement, 20% of fly ash and 10% of silica fume. The fly ash used in the study was obtained from a local coal power plant. The aggregate was a fine sand 0-2 mm. The notation for the mixes are as follows: B – amount of binder, PET – the amount of PET granules in the mix.

The study assumes a constant slump flow of 140 mm ± 10 mm at 15 minutes after adding the water as shown in standard [26]. Similar assumptions for suitability of mixes for 3D printing were proposed in other studies [22, 27–29]. To obtain assumed slump flow value the mixes were modified with superplasticizers. The mix design is presented in Table 1.

**Table 1. Mix composition.**

| No | Mix | Binder [kg] | Cement [kg] | Fly Ash [kg] | Silica fume [kg] | PET [%] | Superplasticizer [kg] |
|---|---|---|---|---|---|---|---|
| 1 | B800/PET0 | 800 | 560 | 160 | 80 | 0% | 1.0 |
| 2 | B800/PET30 | 800 | 560 | 160 | 80 | 30% | 0.7 |

### *2.2. Experimental procedure*

The tests can be divided into two stages: 1) tests conducted on standard samples; 2) tests conducted on 3D printed samples.

The first stage consisted on determination of compressive strength and flexural strength after 24 h and 168 h (7 days) conducted on 4x4x16 cm standard samples [30].

The samples for the second stage of tests were 3D printed with additive manufacturing method. For this purpose, a cartesian robot combined with a concrete extruder was used. Cartesian robot has three axes driven by stepper motors. Everything is controlled by a G-Code. Figure 1 a) presents an example of sample during printing process. Before printing the mixture was evaluated by print quality tests in accordance with [7, 31]. Following tests were conducted on printed samples:

IOP Conf. Series: Materials Science and Engineering 1044 (2021) 012002 doi:10.1088/1757-899X/1044/1/012002

1) Determination of compressive strength and flexural strength vertically to the printing direction (v). A schemes for the tests is presented in Figure 2. The samples were printed in a way to have dimensions as close to standard prisms 40x40x160 mm. The tests were conducted in accordance with [30]. Figure 3 presents the cross sections of B800/PET0 and B800/PET30 after flexural test.
2) Determination of compressive strength of printed structure. A hollow column was designed with R=160 mm ± 5 mm and height of H=205 mm ± 10 mm. The width of the layer was t = 40 ± 5 mm. The column during printing process is shown in Figure 1 a). The view of the column after the test is shown in Figure 1 b).

Mix preparation and printing were made in a laboratory at temperature of 20°C (± 2°) and relative humidity of RH=55% (± 5%).

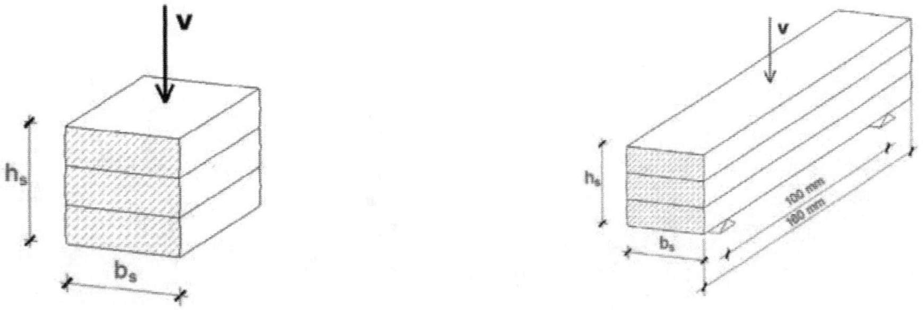


(a) scheme for compression test of printed specimen

(b) scheme for flexural test of printed specimen

**Figure 2.** Schemes for the tests.

(a) B800/PET0 (b) B800/PET30

**Figure 3.** Cross sections of specimens after flexural tests.

## 3. Test results

Figure 4 and Figure 5 presents the comparison of the results of prepared samples after 1 and 7 days. The bulk density of the mix were ρ=2259 kg/m$^3$ for the reference mix B800/PET0 (n=4; CoV=4,95%) and ρ=2001 kg/m3 (n=4, CoV=3,10%) for the modified (alternative) mix B800/PET30.

Figure 4 presents the results of compressive strength determination. Figure 5 presents the results of flexural strength determination. Each test was conducted on at least 4 samples, except of printed columns where 3 specimen were prepared. The value of CoV were presented in the Figure 4.

Based on obtained results it can be said:

1) For standard samples, the compressive strength ratio of samples B800/PET0 to samples B800/PET30 is in the range between 1.21 and 1.28. In the case of flexural strength, the ratio is approximately 1.24. This means that the usage of PET as an aggregate replacement reduced the compressive strength and flexural strength of standard samples after 1 and 7 days by 20-30%.
2) For printed samples, the compressive strength and flexural strength ratio of samples B800/PET0 to B800/PET30 is in the range between 1.04 and 1.10. In this case, the addition of PET also reduced the compressive strength and tensile strength of printed samples. However, the change was lower than for standard prisms.

3) In case of printed columns, the use of PET also decreased the overall strength by approximately 7%. However, the CoV for those samples was much higher (CoV=9%) and the results cannot be taken into consideration without additional tests.
4) The ratio of the compressive strength obtained on the standard specimen to compressive strength obtained on the printed specimen is between 2.48 and 3.27. The similar ratio (standard specimen to printed specimen) for flexural strength is between 1.01 and 1.33. The strength of standard specimen (especially in the case of compression strength) is higher than strength of printed samples. Lower strength of printed samples probably caused by worse compaction and bigger imperfections. Other probable cause lies in interlayer strength in printed samples. Additional tests must be conducted.

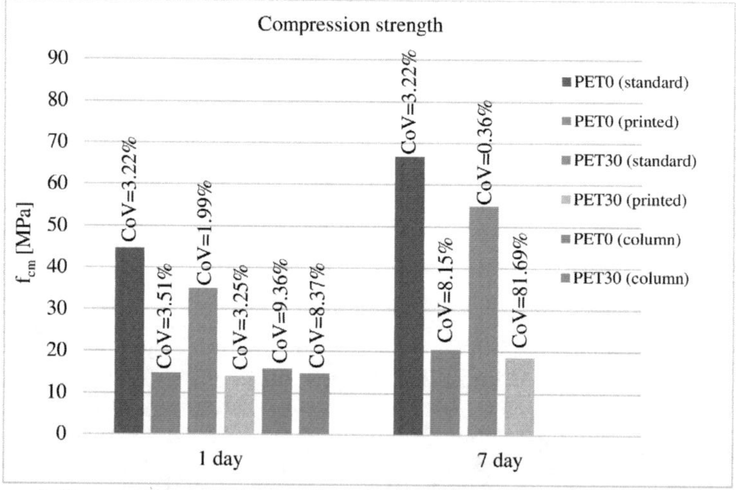


Figure 4. Comparison of the results obtained in compression strength tests.

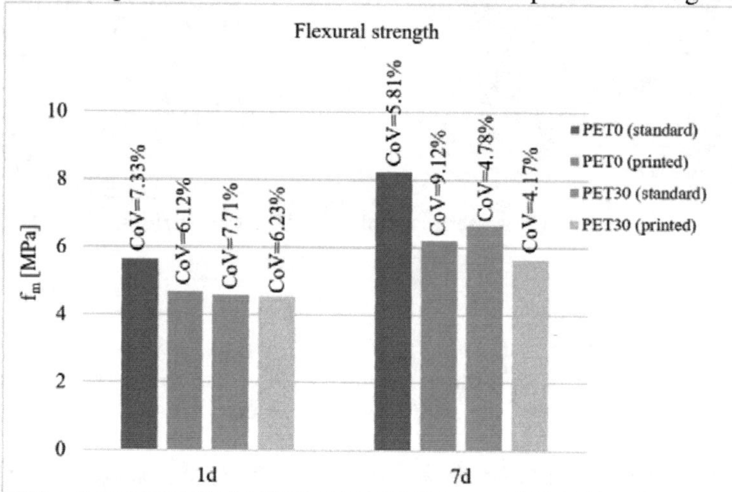


Figure 5. Comparison of the results obtained in flexural strength tests.

## 4. Conclusion

The modification with the PET did not inhibit the required rheological propertied of fresh mix, allowing for proper printing. Additionally, it was possible to pump the mix modified with PET through a typical rotor-stator pumping system, despite increased size of the granules (between 2 mm and 5 mm). To sum up the study proved that the PET granules can be successfully used in 3D printing of cement composites. The study shows that the modification of the mixture with PET reduces its strength by up to 30% in case of standard specimens and by up to 10% in case of printed specimens. However, the most significant conclusion is that even with decreased strength the printing process of column with modified mix was successful, with sufficient strength of the structure. Research shows that mixture with PET additives may become a new direction of research in the field of 3D printing of cement composites. Use of PET in 3D concrete printing technology definitely is line with the idea of sustainable development.

## 5. References

[1] Labonnote N, Rønnquist A, Manum B and Rüther P 2016 Additive construction: State-of-the-art, challenges and opportunities *Automation in Construction* **72** 347–66
[2] Kazemian A, Yuan X, Cochran E and Khoshnevis B 2017 Cementitious materials for construction-scale 3D printing: Laboratory testing of fresh printing mixture *Construction and Building Materials* **145** 639–47
[3] Secrieru E, Fataei S, Schröfl C and Mechtcherine V 2017 Study on concrete pumpability combining different laboratory tools and linkage to rheology *Construction and Building Materials* **144** 451–61
[4] Wangler T *et al* 2016 Digital Concrete: Opportunities and Challenges *RILEM Letters* **1** 67–75
[5] Bos F, Wolfs R, Ahmed Z and Salet T 2016 Additive manufacturing of concrete in construction: Potentials and challenges of 3D concrete printing *Virtual and Physical Prototyping* **11** 209–25
[6] Le T T, Austin S A, Lim S, Buswell R A, Law R, Gibb A and Thorpe T 2012 Hardened properties of high-performance printing concrete *Cement and Concrete Research* **42** 558–66
[7] Kaszynska M, Hoffmann M, Skibicki S, Zielinski A, Techman M, Olczyk N and Wróblewski T 2018 Evaluation of suitability for 3D printing of high performance concretes *MATEC Web Conf.* **163** 1002
[8] Federowicz K, Kaszyńska M, Zieliński A and Hoffmann M 2020 Effect of Curing Methods on Shrinkage Development in 3D-Printed Concrete *Materials* **13** 2590
[9] Kaszynska M and Skibicki S 2017 Influence of Eco-Friendly Mineral Additives on Early Age Compressive Strength and Temperature Development of High-Performance Concrete *IOP Conf. Ser.: Earth Environ. Sci.* **95** 42060
[10] Buswell R A, Soar R C, Gibb A and Thorpe A 2007 Freeform Construction: Mega-scale Rapid Manufacturing for construction *Automation in Construction* **16** 224–31
[11] Ma G, Wang L and Ju Y 2018 State-of-the-art of 3D printing technology of cementitious material—An emerging technique for construction *Sci. China Technol. Sci.* **61** 475–95
[12] Skibicki S 2017 Optimization of Cost of Building with Concrete Slabs Based on the Maturity Method *IOP Conf. Ser.: Mater. Sci. Eng.* **245** 22061
[13] Lowke D, Dini E, Perrot A, Weger D, Gehlen C and Dillenburger B 2018 Particle-bed 3D printing in concrete construction – Possibilities and challenges *Cement and Concrete Research* **112** 50–65
[14] López D, Veenendaal D, Akbarzadeh M and Block P 2014 Prototype of an ultra-thin, concrete vaulted floor system
[15] Siddika A, Mamun M A A, Ferdous W, Saha A K and Alyousef R 2019 3D-printed concrete: Applications, performance, and challenges *Journal of Sustainable Cement-Based Materials* **24** 1–38
[16] Yang H, Chung J K H, Chen Y and Li Y 2018 The cost calculation method of construction 3D printing aligned with internet of things *J Wireless Com Network* **2018** 65
[17] Malaeb Z, Hachem H, Tourbah A, Maalouf T, El Zarwi N. and Hamzeh F. 2015 3d concrete printing: Machine and mix design *International Journal of Civil Engineering and Technology (IJCIET)* **6** 14–22
[18] Panda B, Paul S C, Mohamed N A N, Tay Y W D and Tan M J 2018 Measurement of tensile bond strength of 3D printed geopolymer mortar *Measurement* **113** 108–16
[19] Xia M and Sanjayan J 2016 Method of formulating geopolymer for 3D printing for construction applications *Materials & Design* **110** 382–90
[20] Bong, S.,H., Nematollahi B, Nazari A, Ming Xia and Sanjayan, Jay, G 2019 Fresh and Hardened Properties of 3D Printable Geopolymer Cured in Ambient Temperature *First RILEM International Conference on Concrete and Digital Fabrication -- Digital Concrete 2018*

*(RILEM bookseries, 2211-0852* volume 19*)* ed T Wangler and R Flatt (Cham, Switzerland: Springer)

[21] Lu B, Weng Y, Li M, Qian Y, Leong K F, Tan M J and Qian S 2019 A systematical review of 3D printable cementitious materials *Construction and Building Materials* **207** 477–90

[22] Skibicki S, Kaszyńska M, Wahib N, Techman M, Federowicz K, Zieliński A, Wróblewski T, Olczyk N and Hoffmann M 2020 Properties of Composite Modified with Limestone Powder for 3D Concrete Printing *Second RILEM International Conference on Concrete and Digital Fabrication* ed F P Bos *et al* (Cham: Springer International Publishing) pp 125–34

[23] del Rey Castillo E, Almesfer N, Saggi O and Ingham J M 2020 Light-weight concrete with artificial aggregate manufactured from plastic waste *Construction and Building Materials* **265** 120199

[24] Ismail Z Z and AL-Hashmi E A 2008 Use of waste plastic in concrete mixture as aggregate replacement *Waste Management* **28** 2041–7

[25] Nematollahi B, Vijay P, Sanjayan J, Nazari A, Xia M, Naidu Nerella V and Mechtcherine V 2018 Effect of Polypropylene Fibre Addition on Properties of Geopolymers Made by 3D Printing for Digital Construction *Materials* **11** 2352

[26] EN 1015-3:2000. Methods of test for mortar for masonry - part 3: determination of consistence of fresh mortar (by flow table)

[27] Hoffmann M, Skibicki S, Pankratow P, Zieliński A, Pajor M and Techman M 2020 Automation in the Construction of a 3D-Printed Concrete Wall with the Use of a Lintel Gripper *Materials* **13** 1800

[28] Cho S, Kruger J, Bester F, van den Heever M, van Rooyen A and van Zijl G 2020 A Compendious Rheo-Mechanical Test for Printability Assessment of 3D Printable Concrete *Second RILEM International Conference on Concrete and Digital Fabrication* ed F P Bos *et al* (Cham: Springer International Publishing) pp 196–205

[29] Casagrande L, Esposito L, Menna C, Asprone D and Auricchio F 2020 Effect of testing procedures on buildability properties of 3D-printable concrete *Construction and Building Materials* **245** 118286

[30] EN 1015-11:2020. Methods of test for mortar for masonry - Part 11: Determination of flexural and compressive strength of hardened mortar

[31] Le T T, Austin S A, Lim S, Buswell R A, Gibb A G F and Thorpe T 2012 Mix design and fresh properties for high-performance *Materials and Structures* **45** 1221–32

# Effect of Styrene-Butadiene-Styrene on the Properties and Grading of Local Asphalt Binder in the UAE

**Ali Alnaqbi [1,*], Waleed Zeiada[1,2] Ghazi Al-Khateeb[1,3], Helal Ezzat[1], Abdallah Shanableh[1]**

1.Department of Civil and Environmental Engineering, College of Engineering, University of Sharjah, Sharjah, 27272 Sharjah, UAE.
2.Department of Public Works Engineering, College of Engineering, Mansoura University, Egypt.
3.Jordan University of Science and Technology, Irbid, Jordan

*Corresponding author: U18200514@sharjah.ac.ae

**Abstract.** Flexible pavements in the UAE are subjected to heavy traffic loading repetitions and severe weather conditions which are the main causes of local pavement deteriorate. Fatigue cracking and rutting are most common deterioration types. One effective solution to mitigate both pavement distresses is to use asphalt modifier technologies to improve the properties of asphalt binders used locally in the UAE. The main goal of this study is investigating the effects of Styrene-Butadiene-Styrene (SBS) on the rheological properties of local asphalt binder as well as its grade based on both penetration and Superpave grading systems. Three different SBS contents of 2.5%, 4.5%, and 6.5% by weight of asphalt binder were utilized. It was found that increasing SBS content increased the asphalt binder softening point and rotational viscosity and decreased the penetration value. SBS-modified asphalt binders were less susceptible to temperature changes. Superpave performance results showed that SBS-modified asphalt binders exhibited higher rutting parameter $G^*/\sin\delta$ values as the SBS content increased for both unaged and short-term-aged conditions. Adding 2.5% SBS increased the asphalt penetration grade from 60/70 to 40/50 and the performance grade (PG) from 64 to 70. For the 4.5% and 6.5% SBS cases, the asphalt binder PG increased to 76 and 88 respectively. SBS-modified asphalt binders were found to possess significantly higher mixing and compaction temperature ranges compared to the control asphalt binder.

## 1. Introduction

The traffic volumes in the United Arab Emirates (UAE) had tremendously risen in the last decade due to the fast-pace development of the country. This increase in traffic volumes combined with the harsh environment of the UAE accelerates the deterioration of asphalt pavements especially in areas of congested and heavy traffic [1]. In order to accommodate heavy traffic loadings as well as resist extreme high temperatures, specific polymer-based additives were utilized. Polymer modified asphalt binders have been used successfully at locations of higher loading stresses, such as intersections, airports, vehicle weigh stations, and race tracks [2-6].

There are various types of polymers used to modify and improve the properties of asphalt binders [7-10]. Polymer-modified asphalt binders tend to increase the stiffness at higher temperatures while maintain flexibility at lower temperatures [11], The most commonly used polymer is elastomeric type styrene-butadiene–styrene (SBS) copolymer. The most common form used in asphalt binder

modification is a linear SBS structure [12]. A tremendous number of studies have shown that SBS-modified asphalt binders increased the resistance of asphalt mixtures against rutting [13, 14-17], fatigue cracking [18-20], low-temperature cracking [21-23] and moisture damage [24-25].
When SBS polymer is mixed with asphalt, the polybutadiene phase absorbs the asphalt binder maltene fraction and swells up to nine times its initial volume. At a suitable SBS concentration, normally between 3 and 5%, a polymer network is homogenously formed throughout the asphalt binder matrix and this changes substantially the asphalt binder properties [26-28].

Many researchers have studied the effects of SBS polymers on the asphalt binder properties and they found that the addition of SBS tended to decrease the penetration and increase the softening point of the binder [29-34]. It was found that there is a significant decrease in the penetration values and a considerable increase in the softening point temperatures of SBS-modified asphalt binders at polymer content range of 3% to 5% [35]. For Superpave test, many researchers concluded that increasing SBS content increased the complex shear modulus ($G^*$) and decreased the phase angle ($\delta$). Also, as SBS content increased, rutting parameter ($G^*/\sin\delta$) increased as well, meaning more rutting resistance [31-34, 36, 37].

## 2. Research Objectives

The main objective of this study is to investigate the effect of using different percentages of SBS polymer on the rheological properties and grading of 60/70 local asphalt binder.

## 3. Materials and Testing Plan

All the asphalt binders used in this study were prepared locally in the UAE by MENA Energy Company. The control asphalt binder used in this study is graded as 60/70 according to the penetration grading system. Table 1 presents the basic properties of the 60/70 asphalt binder used locally in most of roadway projects.

**Table 1.** Properties of the control asphalt binder

| Parameter | Value | Standard |
|---|---|---|
| Performance Grade | PG 64−10 | ASTM D 7175 **Error! Reference source not found.** |
| Penetration grade | 60/70 | ASTM D5 **Error! Reference source not found.** |
| Softening point (Ring and Ball) | 47.3°C | ASTM D36 **Error! Reference source not found.** |
| Rotational viscosity at 135°C | 1613 cP | ASTM D4402 **Error! Reference source not found.** |

A total of 4 asphalt binders were tested including the control and SBS-modified asphalt binders using SBS contents of 2.5, 4.5 and 6.5% by weight of asphalt binder. The asphalt binders were tested without aging for penetration system grading. Figure 1 shows the detailed testing plan of this study.

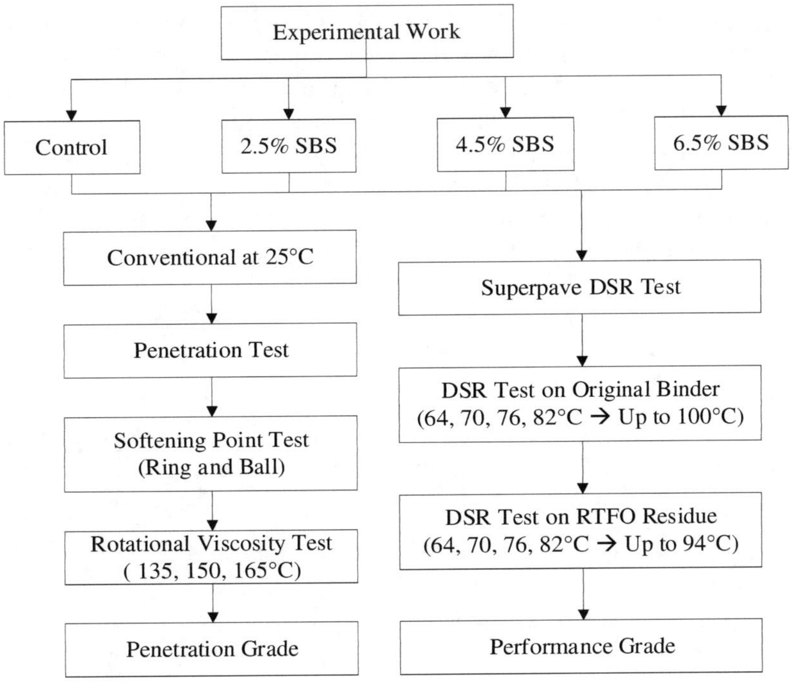


**Figure 1.** Testing plan

## 4. Testing Methods

*4.1 Penetration Test*

According to ASTM D5 [39], the penetration test was carried out to measure the influence of SBS-modification on asphalt binder hardening. First, a set of two standard cans were filled with preheated asphalt binders and were kept at room temperature to cool down for two hours. Afterwards, the samples were immersed in a water bath set at 25 °C degrees for additional two hours. The average of three needle penetrations with 100g applied weight for 5 seconds were then recorded. Two replicate were performed for each asphalt binder.

*4.2 Softening Point Test*

The softening point (ring-and-ball) test was performed based on the ASTM D36 [40] standard to evaluate the softening point of asphalt binder before and after modification. A set of two rings were filled with heated soft asphalt binder then trimmed after cooling for 30 minutes placed on loading frame then immersed into a water bath at a temperature of 5±1 °C. The samples were then heated up in a standard gradual rate under the load of standard steel balls till the asphalt binder samples travels a distance of 25 mm. The measured temperature at this point is set to be the softening point. Two replicate were performed for each asphalt binder.

*4.3 Rotational-Viscometer (RV) Test and Temperature susceptibility*

The rotational viscometer determines the asphalt binder viscosity by measuring the torque necessary to maintain a constant rotational speed of a cylindrical spindle submerged in an asphalt binder specimen held at a constant temperature, as per the ASTM D4402 / D4402M-15 [41]. The Brookfield rotational viscosity values are measured at three temperatures of 135, 150 and 165°C. Standard steel tube is filled with 10 gm of asphalt binder then inserted into the device's chamber which is capable of maintaining a set temperature. The average of three viscosity readings within three minutes was recorded for each temperature after 35 minutes of conditioning [41]. Two replicate were performed for each asphalt binder.

### *4.4 High Temperature Performance Grading*
For the performance grading (PG) system, 25-mm samples were prepared in silicon molds then tested in the parallel plate system of the DSR at max pavement temperatures (64, 70, 76 … etc.) till the bidder reached the 1 kPa limit under a strain level of 0.1 and frequency of 10 Hz. A conditioning of 25 minutes was applied before recording the complex modulus (G*) and the phase angle ( δ ) at a desired temperature [38]. The samples were then short-term aged using rolling thin film oven (RTFO) by placing 8 bottles that have 35 gm of asphalt binder in each inside a disk rotating in the vertical direction at a temperature of 163 °C for 85 minutes subjected to a standard air flow. The RTFO-aged samples were then tested using the DSR 25mm plate as mentioned previously but with a rutting limit of 2.2 kPa [38].

## 5. Results and Discussion

### *5.1 Penetration Test*
The penetration test results are plotted in Figure 2 (a) that illustrates the average penetration value of the two replicate against the SBS content and it is observed that the penetration values decrease as the percentage of SBS increases. Similar results were reported elsewhere in previous studies [29–**Error! Reference source not found..**

### *5.2. Softening Point Test*
The softening point results are plotted in Figure 2 (b). The average softening point temperatures of the two replicate is shown against the SBS content. It is noticed that the softening point temperature of the SBS-modified asphalt binders is higher when compared to that of the control asphalt binder. Furthermore, it is observed that the softening point temperature increases as the SBS content increases. The increase in softening point is an indicator of the stiffening effect of SBS-modified asphalt binder, which is favourable since asphalt binders with higher softening point may be less susceptible to permanent deformation or rutting [29-34].

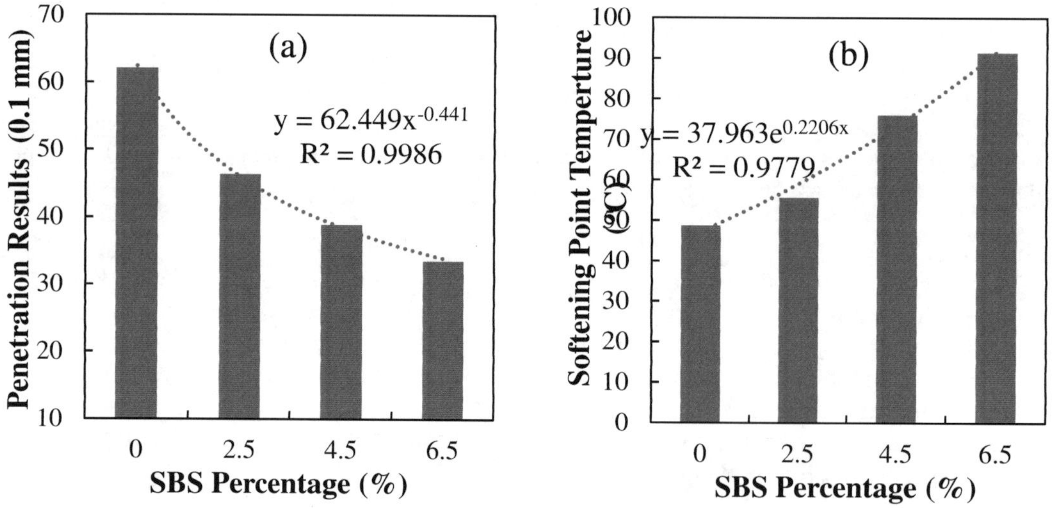


**Figure 2.** (a) Penetration; and (b) softening point test results

### *5.3. Rotational-Viscometer (RV) Test and Temperature susceptibility*
It is generally observed from the RV test results that the viscosity values of all tested asphalt binders were consistently reduced as the test temperatures increases, as shown in Figure 3. It is widely known that the addition of SBS stiffens the asphalt binders [30-32, 34].

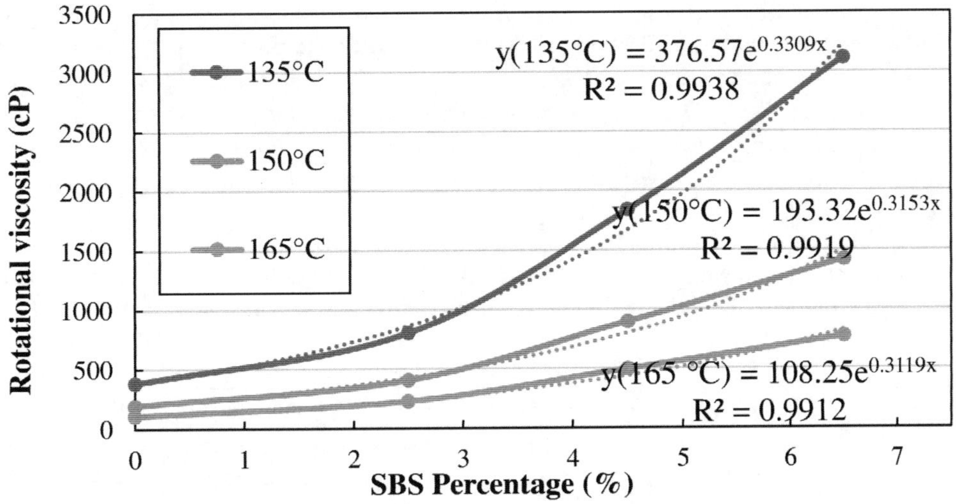


**Figure 3.** Rotational viscosity results at different temperatures

*5.4. High Temperature Performance Grading*

Figure 4 presents the DSR test results of the tested asphalt binders at various temperatures. The testing started typically at 64 °C and then the temperature was increased until the failure condition was attained achieve the minimum requirements of ($G^*/\sin\delta$ = 1 kPa). SBS-modifier is found to significantly improve the performance of asphalt binders indicated by increased $G^*$ and $\delta$ values. These trends of $G^*$ and $\delta$ are known to be favored for the rutting resistance of asphalt binders and asphalt mixtures. The rutting parameter ($G^*/\sin\delta$) of all SBS-modified asphalt binder is higher than this of control asphalt binder at all test temperatures. The rutting resistance of the binder increases with an increase in the percentage of SBS content. Also, by the addition of 2.5, 4.5 and 6.5% SBS to the control asphalt, the rutting resistance is increased by more 2.0, 5.5 and 7.5 times at 64°C, respectively. This indicates an increased elastic behavior with higher SBS content.

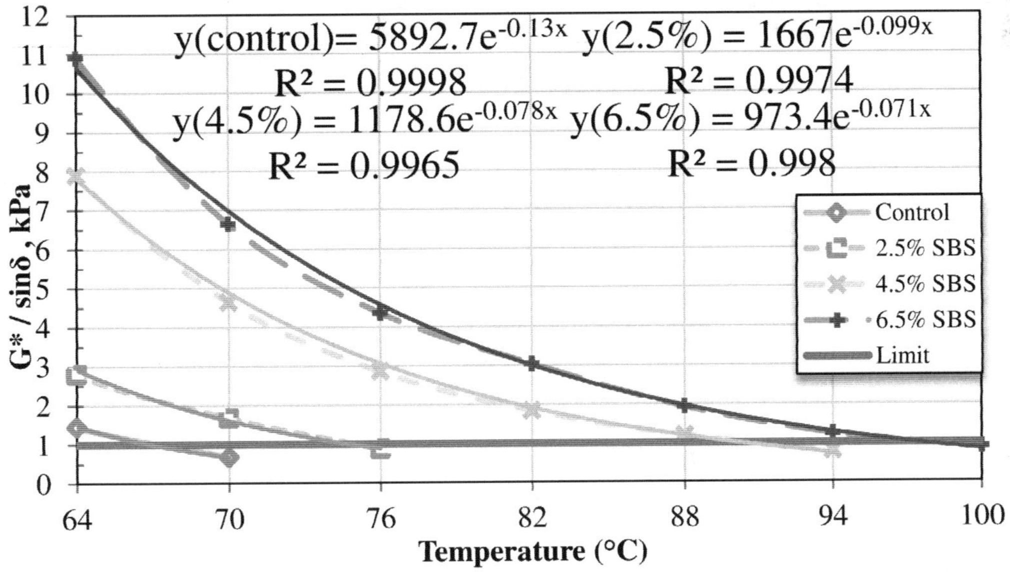


**Figure 4**. DSR rutting parameter ($G^*/\sin\delta$) test results for the un-aged and SBS modified asphalt binders

The results of the DSR tests on RTFO-aged asphalt binders at high temperatures are presented in Figure 5. Figure 5 shows that the SBS-modification successfully improved the rheological properties of the asphalt binders against rutting resistance by increasing G* and decreasing δ. The findings from the DSR test results using RTFO aging condition confirm those obtained from the unaged DSR tests results. The SBS successfully increased the values of PG rutting parameter (G*/sin δ), which is an indication of being more rutting resistant. It is also worth noting that increasing the SBS content is found to increase the rutting parameter. The addition of 2.5, 4.5 and 6.5% SBS to the control asphalt binder increases the rutting resistance by more than 2, 2.8 and 4.3 times at 64°C, respectively. SBS-modified asphalt binders can resist the adverse effects of aging and deformation to a higher extent than the control asphalt binders. This observation confirms the findings from previous studies [31-34, 36-37].

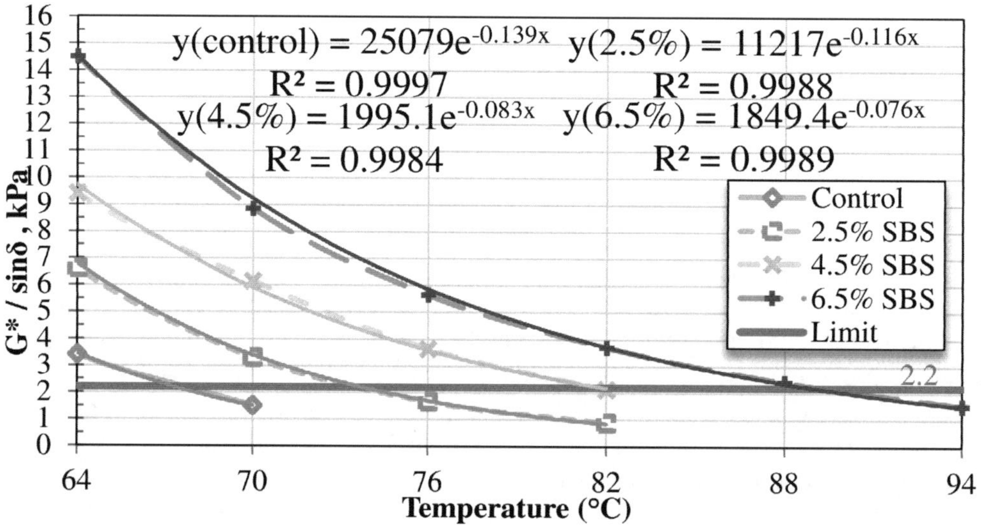


**Figure 5.** DSR rutting parameter (G*/sin δ) test results for the RTFO residues

The effects of temperature on the rutting factor of control and SBS-modified asphalt binders could be represented by the exponential function, expressed as follows:

$$\frac{G^*}{\sin\delta} = \alpha e^{\beta T} \tag{1}$$

Where G*/sin δ is the rutting factor, kPa; $\alpha$ is the properties of asphalt binder; $\beta$ is the influence coefficient of temperature; $T$ is the test temperature, ° C.
According to the test results of control and SBS-modified asphalt binders at different temperatures, we can obtain the model parameters of the regression models as shown in Table 2.

**Table 2.** Summary of regression parameters of control and SBS-modified asphalt binders.

| **Asphalt Binders** | | **Regression parameters** $\alpha$ | $\beta$ | $R^2$ |
|---|---|---|---|---|
| **Un-aged Samples** | Control | 5892.7 | -0.13 | 0.9998 |
| | 2.5% SBS | 1667 | -0.099 | 0.9974 |
| | 4.5% SBS | 1178.6 | -0.078 | 0.9965 |
| | 6.5% SBS | 973.4 | -0.071 | 0.998 |
| **RTFO-aged Samples** | Control | 25079 | -0.139 | 0.9997 |
| | 2.5% SBS | 11217 | -0.116 | 0.9988 |
| | 4.5% SBS | 1995.1 | -0.083 | 0.9984 |
| | 6.5% SBS | 1849.4 | -0.076 | 0.9989 |

*5.5. High Temperature Performance Grading*

Penetration grading classifies different asphalt binders to 5 grades according to ASTM D946 [42] and to 12 grades according to European standards BS EN12591 [43] based on the results of the penetration test at 25 °C. The asphalt binder then has to comply with the limits of other tests or parameters. Table 3 presents the control and the SBS-modified asphalt binders grades as well as the standards limits for both ASTM and BS-EN.

**Table 3**. Summary of asphalt binders penetration grading based on ASTM and BS Standards

| Asphalt Binder | Control | 2.5% SBS | 4.5% SBS | 6.5% SBS |
|---|---|---|---|---|
| **ASTM Grade** | 60/70 | 40/50 | -- | -- |
| **BS Grade** | 50/70 | 40/60 | 35/50 | 30/45 |
| **Penetration at 25 °C (mm)** | 62.10 | 46.33 | 38.83 | 33.50 |
| **ASTM Limits** | 60 - 70 | 40 -50 | -- | -- |
| **BS Limits** | 50 - 70 | 40 -60 | 35 - 50 | 30 - 45 |
| **Softening point (°C)** | 48.7 | 55.6 | 76.1 | 91.5 |
| **ASTM Limits** | > 46 | > 49 | -- | -- |
| **BS Limits** | 46 - 54 | 48 - 56 | 50 - 58 | 52 - 60 |
| **Viscosity at 135 °C ($mm^2/s$)** | 382 | 803 | 1836 | 3112 |
| **ASTM Limits** | -- | -- | -- | -- |
| **BS Limits** | > 295 | > 325 | > 370 | > 400 |
| **Penetration Index (PI) [38]** | -0.975 | -0.087 | 3.180 | 4.524 |
| **ASTM Limits** | -- | -- | -- | -- |
| **BS Limits** | -1.5 - +0.7 | -1.5 - +0.7 | -1.5 - +0.7 | -1.5 - +0.7 |

It can be seen in Table 3 that control and 2.5% SBS-modified asphalt binders achieved the ASTM requirements for 60/70 and 40/50 and the BS standards requirements for 50/70 and 40/60 respectively. There is no grading for the 4.5% SBS and 6.5% SBS according to ASTM. Both of them achieved the requirements for viscosity at 135 °C and PI but failed to comply with the softening point upper limits.

*5.6. Performance grading system*

Superpave grading system classifies asphalt binders to 7 high temperature grades according to AASHTO M320 with several low temperature grades for each making them a total of 37 performance grades. The high temperature grading is based on the DSR ($G^*/\sin\delta$) for both original and RTFO residue. While the low temperature grades are based on the bending beam rheometer (BBR) stiffens (S) and m-value. Table 4 presents the grading of the control and SBS-modified asphalt binders based on AASHTO M320 [44].

According to Table 4, the performance grade (PG) for the control asphalt, 2.5%, 4.5% and 6.5% SBS are finalized as PG 64-XX, PG 70-XX, PG 76-XX and PG 88-XX respectively. Moreover, all the Asphalt binders achieved the viscosity requirements of maximum 3 Ps.s except 6.5% SBS-modified asphalt binder.

**Table 4.** Summary of asphalt binder performance grading based on AASHTO

| Asphalt Binder | Control | 2.5% SBS | 4.5% SBS | 6.5% SBS |
|---|---|---|---|---|
| **Superpave High Performance Grade** | 64 | 70 | 76 | 88 |
| **Original DSR G*/sinδ (kPa)** | 1.45 | 1.67 | 2.85 | 1.93 |
| **AASHTO Limits** | 1.00 | 1.00 | 1.00 | 1.00 |
| **RTFO Residue DSR G*/sinδ (kPa)** | 3.41 | 3.29 | 3.63 | 2.40 |
| **AASHTO Limits** | 2.20 | 2.20 | 2.20 | 2.20 |
| **Viscosity at 135 °C (Pa.s)** | 0.38 | 0.80 | 1.84 | 3.11 |
| **AASHTO Limits** | 3.00 | 3.00 | 3.00 | 3.00 |

### *5.7. Asphalt Binder Workability*

Workability of an asphalt binder is a temperature range that can achieve the viscosity requirements for mixing and handling or laydown and compaction. According to ASTM D2493 [45] the mixing range is (0.17±0.02 Pa.s) and the compaction range is (0.28±0.03 Pa.s). Figure 6 presents the mixing and compaction ranges for all the tested asphalt binders. Figure 6 includes the viscosity verses temperature curves for all tested asphalt binders with mixing and compaction ranges.

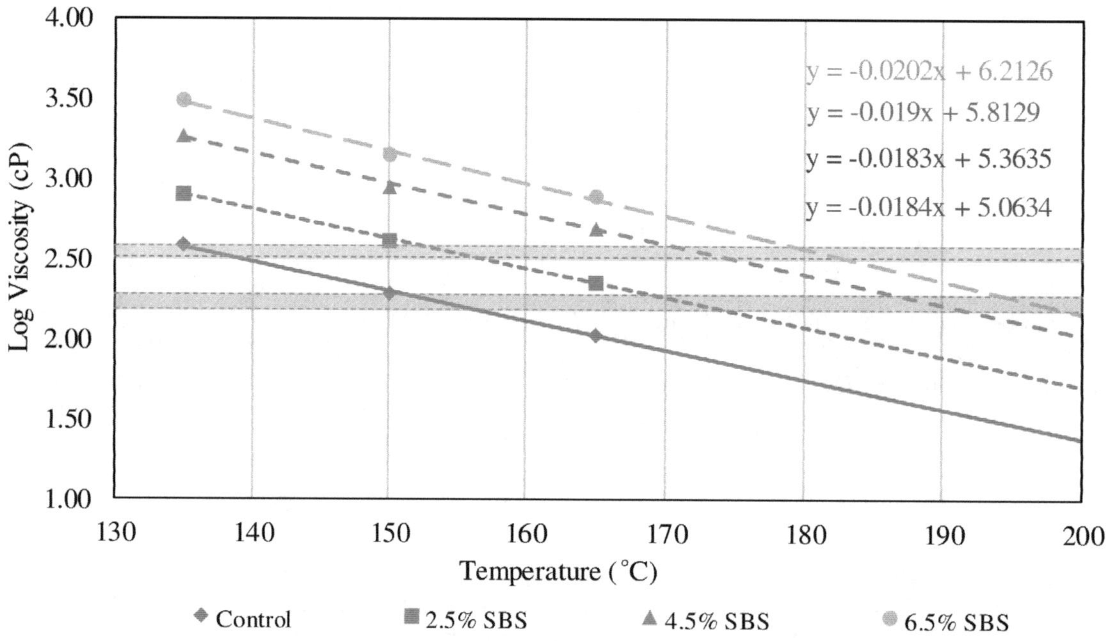


**Figure 6.** Mixing and compaction ranges for all the asphalt binders

Table 5 shows the mixing range, compaction range, mixing workability, delivery limit and compaction workability for control and SBS-modified asphalt binder. Mixing workability is the difference between the upper and the lower limits of the mixing range while the compaction workability is difference between the upper and the lower limits of the compaction range. As shown in Table 5, SBS-modified asphalt binder exhibited a high mixing and compaction ranges. In general, the asphalt binder that has a higher viscosity, a higher softening point, better temperature stability, or a more sufficient viscosity at high temperature can result in a mix that will maintain a certain level of strength and stiffness without a significant shear deformation at high temperature.

**Table 5.** Investigated asphalt binders mixing and compaction ranges

| asphalt binder | Mixing Range | | Compaction Range | | Mixing Workability | Delivery Limit | Compaction Workability |
|---|---|---|---|---|---|---|---|
| **Control** | 156.92 | 151.33 | 139.04 | 134.97 | 5.59 | 12.29 | 4.07 |
| **2.5% SBS** | 174.18 | 168.55 | 156.20 | 152.10 | 5.63 | 12.35 | 4.10 |
| **4.5% SBS** | 191.42 | 185.99 | 174.10 | 170.15 | 5.43 | 11.89 | 3.95 |
| **6.5% SBS** | 199.83 | 194.73 | 183.54 | 179.83 | 5.10 | 11.19 | 3.71 |

## 6. Conclusion

This paper investigated the effect of using SBS at 2.5, 4.5 and 6.5% by weight of asphalt binder on its properties and grading according to the penetration grading system and the Superpave asphalt binder classification system. The testing plan included the penetration, softening point, RV, and DSR tests. The following conclusions can be drawn based on the analysis of the test results obtained for the control and SBS-modified asphalt binders:

•Increasing the SBS polymer content has led to lower penetration values and higher softening point and RV values.

•As the SBS polymer content increased, the $G^*/\sin\delta$ values increased for both unaged and RTFO-aged SBS-modified asphalt binders.

•According to ASTM D946 grading system, 2.5% SBS polymer changed the asphalt binder penetration grade from 60/70 to 40/50 while achieving the softening point, RV and PI requirements.

•According to BS EN12591, 2.5% changed the asphalt binder grade from 50/70 to 40/60 while achieving the requirements of that grade. 4.5% and 6.5% SBS changed the grade to 35/50 and 30/45 respectively but they failed to achieve the softening point requirements.

• Adding SBS polymer contents of 2.5%, 4.5% and 6.5%, changed the high temperature PG grade of the asphalt binder from 64 to 70, 76 and 88 respectively. Only 6.5% SBS failed to comply with the maximum viscosity requirements.

•The mixing and compaction ranges increased with increasing SBS content. The ranges increased up to 45°C more compared to the control asphalt binder when 6.5% SBS was added. The mix design of asphalt concrete (AC) mixtures using SBS-modified asphalt binders to overcome expected workability issues that might be encountered due to increased compaction temperature.

•Further research should be done to investigate the effect of adding SBS on the low temperature PG and to assess the fatigue performance of SBS-modified asphalt binders.

**Acknowledgement**

The authors would like to thank Sharjah research academy for sponsoring this project. The authors would acknowledge the effort of Eng. Ayat Gamal for helping out in some of the penetration tests. Special thanks also to Dr. Hanqi for his help. Thanks to Eng.Leticia Rodriguez from Mena Energy for providing the asphalt binder samples.

## 7. References

[1] Sabouni, R., & Alghazali, A. (2014). Polymer modified asphalt binder asphalt, a proposed solution for UAE pavement deterioration. Construction Materials and Structures, September 2016, 620–628.

[2] King G et al. Additives in asphalt. J Assoc Asphalt Paving Technol A 1999;68:32–69.

[3] Yousefi A. Rubber-polyethylene modified mitumens. Iran Polym J 2004;12:101–12.

[4] Ahmedzade P, Yilmaz M. Effect of polyester resin additive on the properties of asphalt binders and mixtures. Constr Build Mater 2008;22:481–6.
[5] Gorkem C, Sengoz B. Predicting stripping and moisture induced damage of asphalt concrete prepared with polymer modified asphalt binder and hydrated lime. Constr Build Mater 2009;23(6):2227–36.
[6] Airey GD. Rheological evaluation of ethylene vinyl acetate polymer modified asphalt binders. Constr Build Mater 2002;16:473–87.
[7] Brovelli, C., M. Crispino, J. Pais, and P. Pereira. 2015. "Using polymers to improve the rutting resistance of asphalt concrete." Constr. Build. Mater. 77 (2): 117–123.
[8] Cardone, F., G. Ferrotti, F. Frigio, and F. Canestrari. 2014. "Influence of polymer modification on asphalt binder dynamic and steady flow viscosities." Constr. Build. Mater. 71 (11): 435–443.
[9] Yildirim, Y. 2007. "Polymer modified asphalt binders."Constr. Build. Mater. 21(1): 66-72.
[10] Zhu, J., B. Birgisson, and N. Kringos. 2014. "Polymer modification of asphalt binder: Advances and challenges." Eur. Polym. J. 54 (5): 18–38.
[11] John, R., and D. Whiteoak. 2003. Shell asphalt binder handbook. London, UK: Thomas Telford.
[12] Ahmedzade, P. (2013). The investigation and comparison effects of SBS and SBS with new reactive terpolymer on the rheological properties of asphalt binder. Construction and Building Materials, 38, 285–291.
[13] A. Behnood, A. Shah, R.S. McDaniel, M. Beeson, J. Olek, High temperature properties of asphalt binders: Comparison of multiple stress creep recovery and performance grading systems, Transp. Res. Rec. J. Transp. Res. Board. 2574 (2016) 131–143
[14] B.V. Kök, H. Çolak, Laboratory comparison of the crumb-rubber and SBS modified asphalt binder and hot mix asphalt, Constr. Build. Mater. 25 (8) (2011) 3204–3212.
[15] S. Tayfur, H. Ozen, A. Aksoy, Investigation of rutting performance of asphalt mixtures containing polymer modifiers, Constr. Build. Mater. 21 (2) (2007) 328–337.
[16] A. Mokhtari, F. Moghadas Nejad, Mechanistic approach for fiber and polymer modified SMA mixtures, Constr. Build. Mater. 36 (2012) 381–390.
[17] S. Anjan kumar, A. Veeraragavan, Dynamic mechanical characterization of asphalt concrete mixes with modified asphalt binders, Mater. Sci. Eng. A. 528 (21) (2011) 6445–6454.
[18] H. Aglan, A. Othman, L. Figueroa, R. Rollings, Effect of styrene-butadiene styrene block copolymer on fatigue crack propagation behavior of asphalt concrete mixtures, Transp.Res. Rec. 1417. 1417 (1993) 178–186.
[19] R. Lundström, U. Isacsson, Linear viscoelastic and fatigue characteristics of styrene– butadiene–styrene modified asphalt mixtures, J. Mater. Civ. Eng. 16 (6) (2004) 629–638.
[20] T.W. Kim, J. Baek, H.J. Lee, J.Y. Choi, Fatigue performance evaluation of SBS modified mastic asphalt mixtures, Constr. Build. Mater. 48 (2013) 908–916.
[21] K.W. Kim, S.J. Kweon, Y.S. Doh, T.-S. Park, Fracture toughness of polymer-modified asphalt concrete at low temperatures, Can. J. Civ. Eng. 30 (2) (2003) 406–413.
[22] S. Ho, L. Zanzotto, The low temperature properties of conventional and modified asphaltbinders evaluated by the failure energy and secant modulus from direct tension tests, Mater. Struct. 38 (1) (2005) 137–143.
[23] B. Kök, M. Yilmaz, A. Geçkil, Evaluation of low-temperature and elastic properties of crumb rubber– and SBS-modified asphalt binder and mixtures, J. Mater. Civ. Eng. 25 (2) (2013) 257–265.
[24] E. Iskender, A. Aksoy, H. Ozen, Indirect performance comparison for styrene–butadiene– styrene polymer and fatty amine anti-strip modified asphalt mixtures, Constr. Build. Mater. 30 (2012) 117–124.
[25] C.E. Sengul, S. Oruc, E. Iskender, A. Aksoy, Evaluation of SBS modified stone mastic asphalt pavement performance, Constr. Build. Mater. 41 (2013) 777–783.
[26] Yildirim, Y. (2007). Polymer modified asphalt binders. Construction and Building Materials, 21(1), 66–72.
[27] Bru^le´, B.; Brion, Y.; Tanguy, A. Asphalt Paving Technology: Proceedings. Association of Asphalt Paving Technologists, Technical Sessions, 1988, 57, 41

[28] Airey, G. (2003). Rheological properties of styrene butadiene styrene polymer modified road bitumens⋆. Fuel, 82(14), 1709–1719.
[29] Sengoz, B., & Isikyakar, G. Analysis of styrene-butadiene-styrene polymer modified asphalt binder using fluorescent microscopy and conventional test methods. Journal of Hazardous Materials, 150(2)(2008) 424–432.
[30] Sengoz, B., & Isikyakar, G. Evaluation of the properties and microstructure of SBS and EVA polymer modified asphalt binder. Construction and Building Materials, 22(9) (2008) 1897–1905.
[31] Hassanpour-Kasanagh, S., Ahmedzade, P., Fainleib, A. M., & Behnood, A. Rheological properties of asphalt binders modified with recycled materials: A comparison with Styrene-Butadiene-Styrene (SBS). Construction and Building Materials, 230(2020)117047.
[32] Airey, G. D. Rheological properties of styrene butadiene styrene polymer modified road asphalt binders. Fuel, 82(14) (2003)1709–1719.
[33] Gama, D. A., Rosa, J. M., De Melo, T. J. A., & Rodrigues, J. K. G. Rheological studies of asphalt modified with elastomeric polymer. Construction and Building Materials, 106, (2016)290–295.
[34] Ahmedzade, P. The investigation and comparison effects of SBS and SBS with new reactive terpolymer on the rheological properties of asphalt binder Position of. Construction and Building Materials, 38,(2013) 285–291.
[35] Sengoz, B., & Isikyakar, G. (2008). Analysis of styrene-butadiene-styrene polymer modified asphalt binder using fluorescent microscopy and conventional test methods. Journal of Hazardous Materials, 150(2), 424–432.
[36] Behnood, A., & Olek, J. Rheological properties of asphalt binders modified with styrene-butadiene-styrene (SBS), ground tire rubber (GTR), or polyphosphoric acid (PPA). Construction and Building Materials, 151,(2017) 464–478.
[37] Mielczarek M, Słowik M, Andrzejczak K. The assessment of influence of styrene-butadiene-styrene elastomer's content on the functional properties of asphalt binders. Eksploatacja i Niezawodnosc – Maintenance and Reliability 2020; 22 (1)(2020) 148–153,
[38] ASTM D7175-08, Standard Test Method for Determining the Rheological Properties of Asphalt Binder Using a Dynamic Shear Rheometer, ASTM International, West Conshohocken, PA. 2008.
[39] ASTM D5-06, Standard Test Method for Penetration of Bituminous Materials, ASTM International, West Conshohocken, PA. 2006.
[40] ASTM D36-95e1, Standard Test Method for Softening Point of Asphalt binder (Ring-and-Ball Apparatus), ASTM International, West Conshohocken, PA. 2001.
[41] ASTM D4402 / D4402M-15, Standard Test Method for Viscosity Determination of Asphalt at Elevated Temperatures Using a Rotational Viscometer, ASTM International, West Conshohocken, PA. 2015.
[42] ASTM D946 / D946M-20, Standard Specification for Penetration-Graded Asphalt Binder for Use in Pavement Construction, ASTM International, West Conshohocken, PA, 2020, www.astm.org
[43] BS EN 12591:2009 Asphalt binder and bituminous binders - specifications for paving grade asphalt binders (incorporating corrigendum June 2011)
[44] AASHTO M 320-17 Standard Specification for Performance-Graded Asphalt Binder STANDARD by American Association of State and Highway Transportation Officials, 2017
[45] ASTM D2493 / D2493M-16, Standard Practice for Viscosity-Temperature Chart for Asphalt Binders, ASTM International, West Conshohocken, PA, 2016, www.astm.org

# The Water-Dependent Evolution of Pore Size Distribution due to Hydro-Chemical Transport within Cement-based Materials

**XIONG Qing xiang* and MEFTAH Fekri**

Laboratory of Civil and Mechanical Engineering (LGCGM), National Institute for Applied Sciences, Campus Beaulieu, 35700 Rennes, France

*E-mail: qingxiang.xiong@insa-rennes.fr

**Abstract.** This work explores the evolution of pore structure by considering the coupling effect of ionic transport and chemical reaction, especially salt precipitation. The reactive transport system is based on two sets of equations: a set of partial differential equations based on mass conservation for the ionic transport, and a set of nonlinear algebraic equations based on the law of mass action for the chemical processes. By adopting this model, transport properties such as ionic concentration and mass of chemical components are implemented. The porosity and the water saturation degree are also considered and examined in our work. Besides, we propose a probabilistic-based porous network representing the evolution of microstructure induced by precipitation/dissolution. The probabilistic porous network is the interconnection between the microstructure and the macro properties. With a given water saturation degree and porosity calculated by the reactive transport model, we could establish the water-dependent evolution of pore structure in cement-based materials. Such a numerical model could be used to interpret the mechanism of the local precipitation/dissolution process in pore scales, which cannot be implemented by experimental measurements.

## 1. Introduction

The cement-based material, a kind of heterogeneous porous material, is becoming a dominative application in building construction industry. As the most characterization for the porous material, the pore structure, not only the porosity but also the pore size distribution (PSD), determines the durability of porous material by affecting the mechanical behaviour and its transport properties [1]. On the contrary, owing to the aggressive surroundings, the porous structure is changed by the ingress of contaminants such as water, carbon dioxide, salt and sulphate.

As one of the most crucial criteria to evaluate durability, transport properties which involving chemical reactions are studied in this research. Being different from previous researchers that devote too much to complexation and adsorption reactions, this work focuses only on the precipitation/dissolution reaction over time during the ionic transport process.

Besides, it should be noted that the transport properties are not sufficient to provide an overall criterion for durability. The physical characteristics, such as its pore structure and porosity, are of equal importance. Therefore, herein, we also concern the evolution of PSD induced by the collaboration of ionic transport and chemical reaction within cement-based material.

Correspondingly, much research work has been dedicated to determining the pore structure; among them, mercury intrusion porosimetry (MIP) is a commonly used method because it is quick and convenient [2, 3]. However, the MIP method is not adequate to describe the wide range of pore size for cement-based material [4]. In1929, Hatch [5] presented the early studies of fitting experimental of pore size with lognormal function. After that, Diamond [6], D. Shi [2, 7], Holly [8] developed the

single or several sub-distribution of lognormal distribution to fit the PSD. The results indicated that the lognormal distribution has a physical fundamental in the description of PSD. Thus, we adopt the lognormal distribution to interpret the microstructure of porous material.
To arrive at the correct modelling of hydro-chemical transport process that affects the evolution of pore size distribution, research is carried out on the most common damaging agents for the materials- the salt precipitation/dissolution. In the first part, the reactive transport model is developed based on the mass conservation equation and the law of mass action [9]. Then the proposed probabilistic method is introduced to illustrate the porous structure. The numerical model is then validated by implementing an example, and finally, the conclusions are discussed.

## 2. Methodology

### 2.1. Treatment of hydro-chemical coupling model

Let us propose that in the targeted system, there are a subset of $N_x$ aqueous species and a subset of $M_p$ precipitated species. According to Fick's law and the mass conservation equation, ionic transport can be written as

$$M_j \phi S_w \frac{\partial c_j}{\partial t} + M_j \phi c_j \frac{\partial S_w}{\partial t} + M_j S_w c_j \frac{\partial \phi}{\partial t} + Q_c = 0 \tag{1}$$

where $Q_c = div\left(m_j v_{j-w}\right) + div\left(m_j v_{w-s}\right)$ represents the diffusion of the ionic species, $m_j$ is the relative mass of $j$th ionic species, $v_{j-w}$ and $v_{w-s}$ is the velocity of ionic species and water, respectively, $c_j$ is the concentration of the $j$th ionic species, $M_j$ is the molar mass of the $j$th ionic species, $S_w$ is the water saturation degree and $\phi$ is the total porosity.
The evolution equation for porosity in the process of precipitation/dissolution is given by

$$\frac{\partial \phi}{\partial t} = -\frac{\partial \phi_{salt}}{\partial t} \tag{2}$$

where $\phi_{salt}$ is the pore volume fraction for precipitated salt.

The precipitation/dissolution of solid-phase is the chemical reaction described by a set of nonlinear algebraic equations under the assumption of local equilibrium state. It is governed by the following expression based on the law of mass action [9]:

$$K_{eq} \prod_{j=1}^{N_x} \left(c_j\right)^{a_{ij}} = 1 \qquad i = 1, 2, \ldots, M_p \tag{3}$$

$K_{eq}$ is the thermodynamic equilibrium constant for the $i$th chemical reaction; $a_{ij}$ is the stoichiometric coefficient of the $j$th ionic component in the $i$th precipitated species.

### 2.2. Water-dependent evolution model for pore size distribution

This section focuses on the elaboration of a morphological change model (by precipitation/dissolution) of the PSD by considering the location of the capillary interface.
The model is based on the given law of probability for the initial pore size distribution $P_0\left(r_0\right)$ and a transformation law for the pore radius. The transformation law is aimed to determine the new distribution $P_1\left(r_1\right)$ based on the overall value of the porosity resulting from the transfer model.
After introducing $\phi_0$ and $\phi_1$, respectively, the overall value of the porosity before and after the chemical reaction, we have:

$$\phi_1\left(r_1\right) = \int_0^{r_1} \phi_1 P_1\left(u\right) du \tag{4}$$

The transformation relation of the pore radius could be defined as equation (5), which indicates an initial pore radius $r_0$ becomes a modified pore radius $r_1$ after the solid constituting the skeleton undergoes a deposition (or an erosion) on a distance $\varepsilon(r_0)$ (see figure 1).

$$r_1 = f(r_0) = r_0 + \varepsilon(r_0) = r_0 + \bar{\varepsilon} H(r_0) \tag{5}$$

Where $H(r_0)$ is a symmetric S-shape function $\left(0 \leq H(r_0) \leq 1\right)$. $\bar{\varepsilon}$ is a constant (if salt precipitation occurred: $\bar{\varepsilon} < 0$, if salt dissolution occurred: $\bar{\varepsilon} > 0$). More details can be found in appendices.

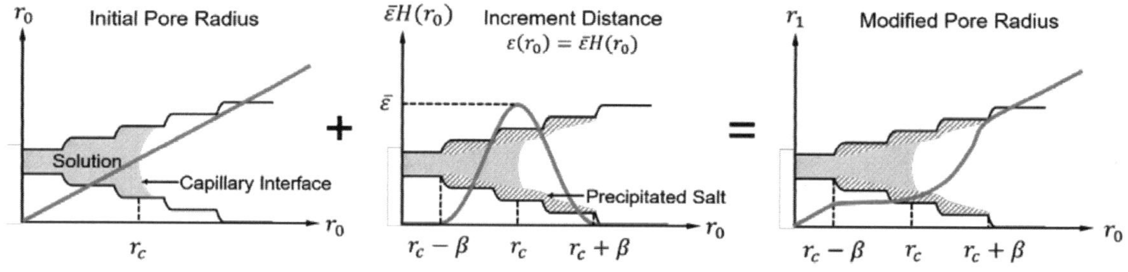


**Figure 1.** Schematic of transformation law for modified pore size ($r_c$ is the location of capillary interface, $\beta$ denotes the shape parameter $(\beta > 0)$).

The pores are considered as cylindrical. Based on the assumption that the ratio of variation for porosity before and after transformation is equal to the ratio of volume variation of the pore of respective radius $r_0$ and $r_1$, we have

$$\frac{d\phi_1}{d\phi_0} = \frac{r_1 dr_1}{r_0 dr_0} \tag{6}$$

We can then deduce the modified probability density function for new PSD as follows:

$$P_1(r_0) = \frac{\phi_1}{\phi_0} \frac{f(r_0)}{r_0} \frac{df(r_0)}{dr_0} P_0(r_0) \tag{7}$$

## 3. Implemented Example

The model deals with time-dependent reactive transport. The most common reaction- precipitation of sodium chloride is studied by giving the initialization parameters as in table 1:

**Table 1.** Initialization of parameters for representing porous solution.

| $C_{Na^+}(mmol/L)$ | $C_{Cl^-}(mmol/L)$ | $\phi_0$ | $S_w$ | $K_{eq}$ | $Q_c(mmol/L.h)$ |
|---|---|---|---|---|---|
| 10 | 10 | 0.2 | 0.4 | 0.01242 | -1 |

In this research, let us specialize it to the case of a masonry cement whose pore size distribution is obtained by a MIP test. The initial pore size distribution $P_0(r_0)$ of the cement-based material is considered as lognormal distribution [2].The fitting parameters of initial PSD is summarized in table 2, where $\mu_r$ is the location parameter, and $\sigma_r$ is the standard deviation.

Figure 2 and 3 give us profiles of each component varies with the ionic transport coupling with the salt precipitation reaction. During this process, the water saturation degree increases from 0.4 to 1, while the porosity decreases from 0.2 to 0.08 (see figure 2). It can be found that ionic diffusion affects the system by means of a pronounced increment in the mass of precipitated salt, which is over several orders of magnitude (see figure 3).

**Table 2.** Tested properties and fitted parameters of the lognormal distribution model.

| $\phi_0$ | $r_{\max}(\mu m)$ | $r_{\min}(\mu m)$ | Cumulative Pore Volume $(mL/g)$ | $\sigma_r$ | $\mu_r$ |
|---|---|---|---|---|---|
| 0.2 | 160 | 0.0015 | 0.0971 | 1.2 | -1.8788 |

At the same time, the porous structure is also affected by the hydro-chemical transport process. Here we work on the period with the porosity decreasing from 0.2 to 0.16 by a step of 0.05 for each time.

Figure 4 shows the evolution of PSD due to reactive transport. The peak value of the probability density increases as the decreased porosity, while the mode value has a slight shift to left side, which indicates that the precipitation process creates more small pores. It is worthy of valuing the PSD develops into bimodal when the porosity decreases to a specific value. This phenomenon reveals that precipitated salt fills partial pores near the waterfront, which divides the large capillary pores into smaller capillary pores. Consequently, the unimodal distribution develops into a bimodal distribution that contains different populations of pore size. This phenomenon was also introduced in D. Shi's work in 1991[7].

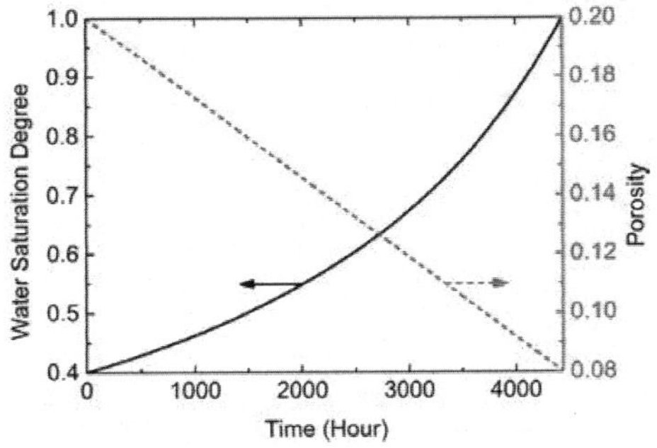


**Figure 2.** Water saturation degree varies with time.

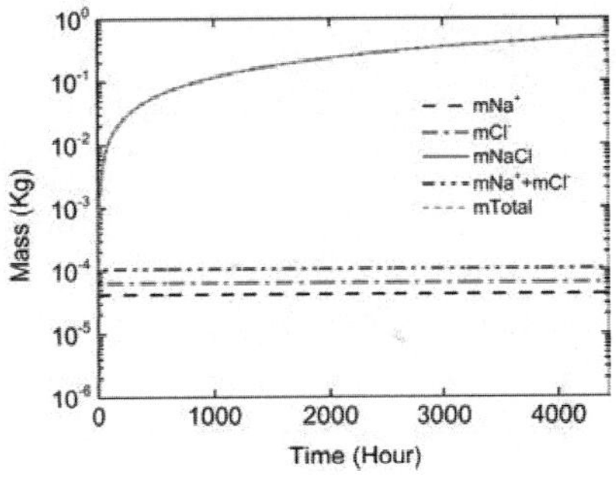


**Figure 3.** Evolution of mass for each component.

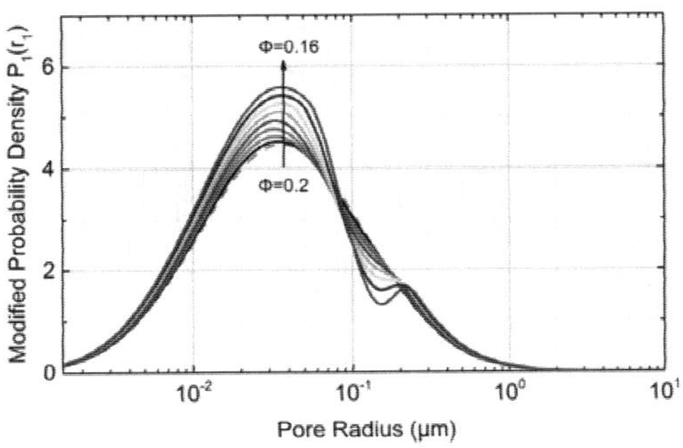


**Figure 4.** Evolution of pore size distribution due to ionic transport and chemical reaction.

## 4. Conclusions

This work developed a probabilistic method to describe the PSD by concerning ionic transport model coupled with salt precipitation reaction. With this numerical model, we could conduct a quantitative

analysis of salt precipitation during the ionic transport process, which provided the possibility to elucidate the evolution of PSD induced by local precipitation/dissolution in the pore scale.
However, this transport system was developed based on a set of hypothesis. Besides, the gaseous phase of this model was negligible, which might lead to ionic transport deviates from reality. For further research, time and spatial dependent transport models should be developed to improve the fully understanding of reactive transport problems.

## 5. Appendices

The symmetric S-shape function $H\left(r_0\right)$ is defined as

$$H\left(r_0\right)=\begin{cases} 0 & 0\le r_0\le r_c-\beta \\ G\left(r_0\right)=\frac{3}{4}\left(\frac{2r_0+\beta-2r_c}{\beta}-\frac{\left(2r_0+\beta-2r_c\right)^3}{3\beta^3}+\frac{1}{2}\right) & r_c-\beta<r_0\le r_c \\ K\left(r_0\right)=-\frac{3}{4}\left(\frac{2r_0-\beta-2r_c}{\beta}-\frac{\left(2r_0-\beta-2r_c\right)^3}{3\beta^3}+\frac{1}{2}\right) & r_c<r_0\le r_c+\beta \\ 0 & r_c+\beta<r_0<\infty \end{cases} \tag{8}$$

As the porosity fraction function after the chemical reaction is given by

$$\phi_1=\int_0^\infty \phi_1 P_1\left(r_1\right)dr_1=\phi_0\left(1+\bar{\varepsilon}I_1+\bar{\varepsilon}^2 I_2\right) \tag{9}$$

The constant $\bar{\varepsilon}$ should be

$$\bar{\varepsilon}=\frac{-I_1\pm\sqrt{I_1^2-4I_2\left(1-\frac{\phi_1}{\phi_0}\right)}}{2I_2} \tag{10}$$

In which $I_1, I_2$ are definite integrations that are expressed as

$$I_1=\int_{r_c-\beta}^{r_c}\left(\frac{G\left(r_0\right)}{r_0}+\frac{dG\left(r_0\right)}{dr_0}\right)P_0\left(r_0\right)dr_0+\int_{r_c}^{r_c+\beta}\left(\frac{K\left(r_0\right)}{r_0}+\frac{dK\left(r_0\right)}{dr_0}\right)P_0\left(r_0\right)dr_0 \tag{11}$$

and

$$I_2=\int_{r_c-\beta}^{r_c}\left(\frac{G\left(r_0\right)}{r_0}\cdot\frac{dG\left(r_0\right)}{dr_0}\right)P_0\left(r_0\right)dr_0+\int_{r_c}^{r_c+\beta}\left(\frac{K\left(r_0\right)}{r_0}\cdot\frac{dK\left(r_0\right)}{dr_0}\right)P_0\left(r_0\right)dr_0 \tag{12}$$

## 6. References


[1] Huang Q, Jiang Z, Gu X, Zhang W and Guo B 2015 Numerical simulation of moisture transport in concrete based on a pore size distribution model *Cement and Concrete Research* **67** 31–43

[2] Shi D, Ma W and Brown P W 1989 Lognormal Simulation of Pore Evolution During Cement and Mortar Hardening *MRS Proc.* **176** 143

[3] Diamond S 1971 A critical comparison of mercury porosimetry and capillary condensation pore size distributions of portland cement pastes *Cement and Concrete Research* **1** 531–45

[4] Roels S, Elsen J, Carmeliet J and Hens H 2001 Characterisation of pore structure by combining mercury porosimetry and micrography *Mat. Struct.* **34** 76–82

[5] Hatch T and Choate S P 1929 Statistical description of the size properties of non uniform particulate substances *Journal of the Franklin Institute* **207** 369–87

[6] Diamond S and Dolch W L 1972 Generalized log-normal distribution of pore sizes in hydrated cement paste *Journal of Colloid and Interface Science* **38** 234–44

[7] Shi D, Brown P W and Ma W 1991 Lognormal Simulation of Pore Size Distributions in Cementitious Materials *Journal of the American Ceramic Society* **74** 1861–7

[8] Holly J, Hampton D and Thomas M D A 1993 Modelling relationships between permeability and cement paste pore microstructures *Cement and Concrete Research* **23** 1317–30
[9] Yeh G-T and Tripathi V S 1991 A Model for Simulating Transport of Reactive Multispecies Components: Model Development and Demonstration Water Resources Research **27** 3075–94

# Accuracy of RSA Design Demands for Multi-Tower Buildings

**Tarek Youssef[1] and Chatpan Chintanapakdee[2,*]**

1 Department of Civil Engineering, Faculty of Engineering, Chulalongkorn University, Bangkok, Thailand
2 Centre of Excellence in Earthquake Engineering and Vibration, Faculty of Engineering, Chulalongkorn University, Bangkok, Thailand

E-mail: chatpan.c@chula.ac.th

**Abstract.** Response spectrum analysis (RSA) procedure commonly used in design practice has been found to underestimate design forces in tall buildings. This study aims to check the accuracy of the RSA procedure when applied to multi-tower buildings sharing a common podium. Two hypothetical multi-tower buildings, the first consists of two towers with the same height (building SH), and the second consists of two towers with the different height (building DH), were designed for Chiang Mai site of Thailand using the conventional RSA procedure, and then the nonlinear response history analysis (NLRHA) is carried out. Results from NLRHA, which is the most accurate method, were used as reference values to evaluate the accuracy of RSA procedure. The results show that seismic shear demand from nonlinear analysis is about 3 times larger than the demand obtained from the RSA procedure. This could lead to shear failure in the shear walls designed by RSA procedure. Since the elastic method is preferred in practice, a modified response spectrum analysis (MRSA) procedure previously proposed for computing shear demand in regular tall buildings, was tried to apply to irregular tall buildings in this study. It is found that MRSA provides good estimates of shear force in vertical elements. MRSA can significantly enhance the design of tall buildings by avoiding brittle shear failure.

## 1. Introduction

Reinforced concrete (RC) shear walls are commonly used as lateral force-resisting system in tall buildings. To design such structures to resist earthquake, several methods would be used such as: equivalent lateral force (ELF), response spectrum analysis (RSA) procedure, and performance-based design (PBD) approach, which requires nonlinear response history analysis (NLRHA). As allowed in ASCE 7-16 [1] the RSA procedure is widely used in current practice to compute design demands of structures. However, there are limitations on height, type and irregularity of structural system that can be used in ASCE 7-16. Structural irregularities are one of the major causes of damage amplification under seismic action as it can increase in seismic demands of a structural system. There are several special considerations in ASCE 7-16, when dealing with irregular tall buildings. For example, for seismic design category (SDC) E, structural system with extreme torsional irregularity, weak story or extreme soft story irregularity is not permitted. Therefore, the dynamic response of irregular structures is an issue that is explicitly warned against in design guidelines. However, design engineers do not pay much attention to those limitations, and they use RSA procedure in the code to compute the design demands of a tall building with RC shear walls with a height taller than 48.8m, which is not compliant

IOP Conf. Series: Materials Science and Engineering 1044 (2021) 012005 doi:10.1088/1757-899X/1044/1/012005

to the scope of the prescriptive code. This can lead to the unsafe design of tall building because RSA procedure has been found to underestimate the actual demands computed from NLRHA procedure.
The issue of underestimation of design demands obtained from the RSA procedure has been noticed by many researchers. There are many reasons for the inaccuracy of the RSA procedure, which are:

1. Shear forces can still increase after flexural yielding occurs [2], while building codes assume that shear forces are limited by flexural yielding and they would reduce by the same manner.
2. The use of a single force response reduction factor ($R$) to reduce the total combined response is one of the primary reasons. Several researchers have demonstrated that higher modes are not significantly affected by inelasticity as much as the first mode [3-6].
3. Flexural overstrength inherent in design is the reason for excessive force demand needed to cause flexural yielding [2, 7]. There are many sources of flexural overstrength such as expected material strength which is greater than the material strength used in the design, reduction factor used in the design, and minimum reinforcement requirement.

Many subsequent studies have proposed different approaches to overcome the issue of underestimation of shear demand obtained from the RSA procedure. Khy et al. [8] conduct a study on RC shear wall tall buildings ranging between 15 to 39 story, subjected to earthquake excitations in Bangkok are first designed using the RSA procedure, and then NLRHA is carried out. It was found that the RSA procedure underestimates shear demand when compared to NLRHA results. Khy et al. [8] proposed a modified response spectrum analysis (MRSA) to compute shear forces in tall buildings, and it was found that MRSA provides good estimates of shear forces in RC shear wall tall buildings. Therefore, MRSA method is tried to apply to irregular tall buildings in this study.
Two versions of MRSA were proposed, the first is based on higher-mode elastic approach denoted as "$\text{MRSA}_{\text{HE}}$", and the second is based on a higher-mode inelastic approach denoted as "$\text{MRSA}_{\text{HI}}$". The shear demand of $\text{MRSA}_{\text{HE}}$ and $\text{MRSA}_{\text{HI}}$ is computed from equation (1) and (2) respectively.

$$V_{\text{HE}} = I \times \sqrt{\left( \frac{SF \times \Omega_0}{R} V_{1e} \right)^2 + V_{2e}^2 + V_{3e}^2 + \ldots} \tag{1}$$

$$V_{\text{HI}} = I \times \sqrt{\left( \frac{V_{1e}}{R_1} \right)^2 + \left( \frac{V_{2e}}{R_2} \right)^2 + \left( \frac{V_{3e}}{R_3} \right)^2 + \ldots} \tag{2}$$

where $SF$ is the scale factor, $\Omega_0$ is the overstrength factor, $I$ is the importance factor, $V_{ie}$ is the shear of the $i^{\text{th}}$ mode.

### 1.1. Objective
The objective of this study is to check the accuracy of conventional RSA and previously proposed MRSA method when applied to irregular tall buildings and to suggest a practical way that can be used for such structure.

## 2. Methodology
The procedure to conduct this study is outlined as the followings:

1. Create the linear mathematical model using ETABS software [9].
2. Compute the design demands that results from all factored load combinations including gravity load, wind load, and seismic forces from RSA procedure.
3. Design the structural system according to ACI 318M-14 [10].
4. Create the nonlinear mathematical model using PERFORM-3D software [11], and analyze the structure by NLRHA.
5. Compute the force response reduction factor ($R_i$) for the selected modes using pushover analysis.
6. Compute the demand forces from both $\text{MRSA}_{\text{HE}}$ and $\text{MRSA}_{\text{HI}}$ methods [8].
7. Evaluate the accuracy of RSA and MRSA methods by comparing them with the results of NLRHA.

IOP Conf. Series: Materials Science and Engineering 1044 (2021) 012005 doi:10.1088/1757-899X/1044/1/012005

## 3. Description of studied buildings and earthquake ground motions

Two irregular RC shear wall tall buildings are designed for Chiang Mai site in Thailand. Both buildings consist of two towers sharing a common podium. One of the buildings consists of two towers with the same height, and it is denoted as building "SH". The other building consists of two towers with different heights, and it is denoted as building "DH". Building SH and DH are shown in Figure 1 and 2 respectively.

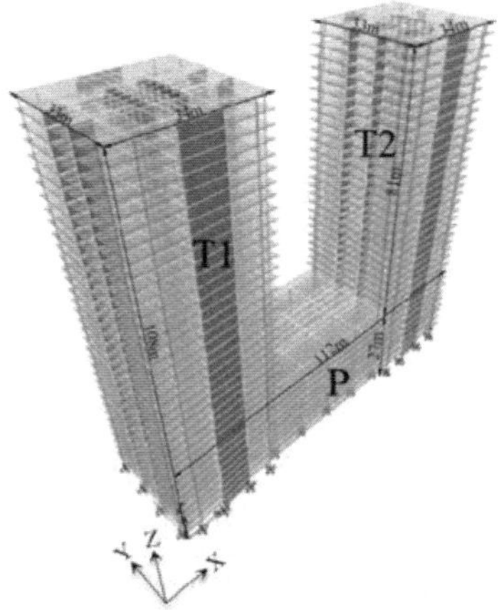


**Figure 1**. Three-dimensional model of building SH.

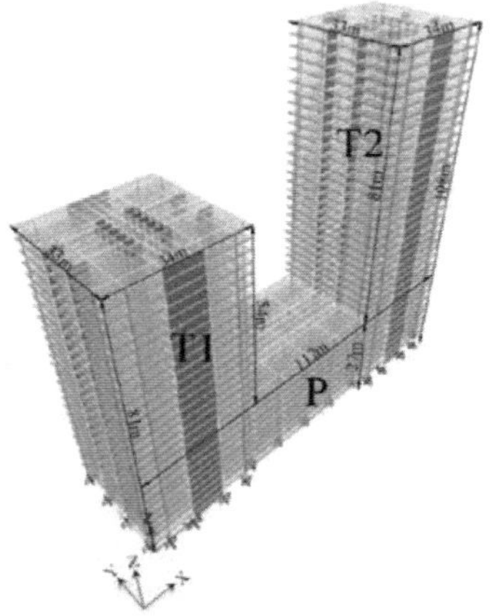


**Figure 2**. Three-dimensional model of building DH.

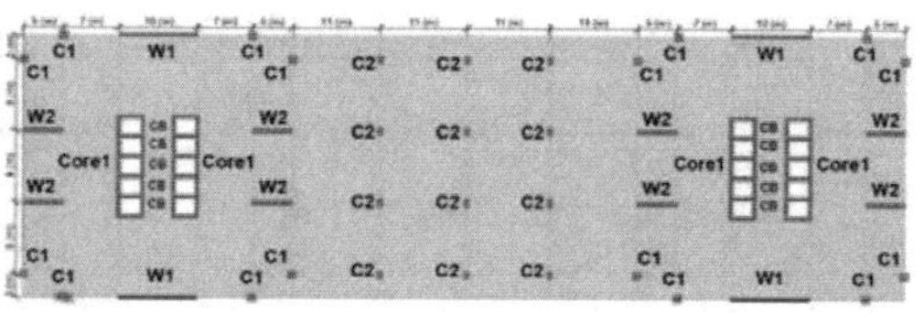

**Figure 3**. Podium floor plans (1$^{st}$ to 9$^{th}$ floor).

**Figure 4**. Tower floor plans (10$^{th}$ to top floor).

The lateral force resisting system was considered to be special RC shear wall whose design factors according to ASCE 7-16 [1] are shown in Table 1, where $R$ is the response modification factor, $C_d$ is the deflection amplification factor, $\Omega_0$ is the overstrength factor, and $I$ is the important factor. Design wind pressure is taken from Bangkok Building Control Act (2001) [12] with the values shown in Table 2.

**Table 1.** Design factors according to ASCE 7-16 [1].

| Factors | Values |
|---|---|
| $R$ | 6 |
| $C_d$ | 5 |
| $\Omega_0$ | 2.5 |
| $I$ | 1.25 |
| Risk category | III |
| Site class | D |
| SDC | D |

**Table 2**. Wind load pressure according to Bangkok Building Control Law Act [12].

| Building height ($H$) | Design Wind Pressure (kN/m$^2$) |
|---|---|
| $H \leq 10$ m | 0.5 |
| $10 < H \leq 20$ m | 0.8 |
| $20 < H \leq 40$ m | 1.2 |
| $40 < H \leq 80$ m | 1.6 |
| $H > 80$ m | 2 |

The in-plane flexural stiffness used in the linear and nonlinear model is shown in Table 3. The out-of-plane behavior of walls and slabs was assumed to remain elastic in both linear and nonlinear model with small effective stiffness of $0.25E_cI_g$.

For the nonlinear model, a nonlinear fiber element was used for RC walls and columns, while moment-rotation hinges were assigned for coupling beams. The nonlinear fiber behavior is controlled by the nonlinear stress-strain relation of concrete and steel material properties. The stress-strain relation of concrete proposed by Mander et al. [13] was adopted. The expected strength was used in the model, as the actual strength is usually greater than the nominal strength specified by the designer. The expected material strength of concrete and steel were taken as 1.25 times the nominal strength.

**Table 3**. Effective stiffness of structural members in linear and nonlinear model.

| | Linear model | | Nonlinear model | |
|---|---|---|---|---|
| **Elements** | **Flexural** | **Axial** | **Flexural** | **Axial** |
| Shear wall | 0.35 | 1.0 $E_cA_g$ | Fiber | Fiber |
| Slab | 1.0 $E_cI_g$ | 1.0 $E_cA_g$ | 1.0 $E_cI_g$ | 1.0 $E_cA_g$ |
| Column | 0.7 $E_cI_g$ | 1.0 $E_cA_g$ | Fiber | Fiber |
| Beam | 0.35 | 1.0 $E_cA_g$ | 0.35 | 1.0 $E_cA_g$ |

where $E_c$ is the modulus of elasticity of concrete; $I_g$ is the gross moment of inertia of cross section; $A_g$ is the gross cross-sectional

Chiang Mai site located on stiff soil (site class D), where the target spectrum (UHS) was developed using the procedure described in ASCE 7-16, for a response spectral parameter of $S_S$ = 0.963g, and $S_1$ = 0.248g.The damping factor formula in ASCE 41-13 [14] was used to obtain the design spectrum curve that corresponds to 2.5% damping. The spectral matching ground accelerations were obtained using ten ground motions that have similar seismic mechanism of the studied site [15]. The individual matching spectra, the mean value of matching spectra, and the target spectrum (UHS) for 2.5% damping ratio in Chiang Mai are shown in figure 5.

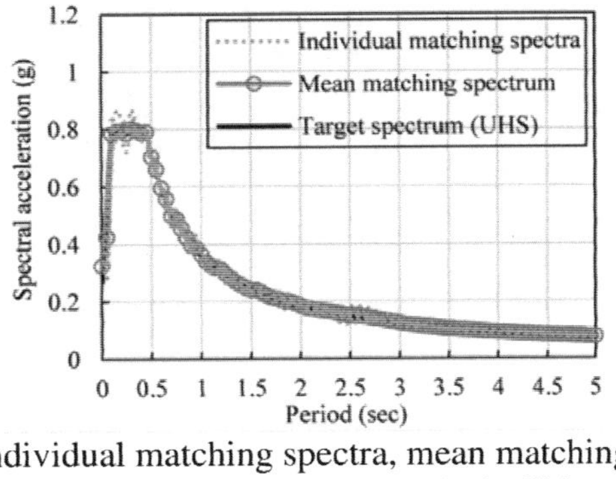


**Figure 5**. Individual matching spectra, mean matching spectrum, and target spectrum for 2.5% damping ratio in Chiang Mai.

## 4. Response spectrum analysis procedure

The concept of RSA procedure described in building codes is to reduce the elastic demand by response modification factor (*R*) and to amplify the displacements by deflection amplification factor ($C_d$), as shown in equations (3) to (5). The scaling factor (*SF*) is computed according to ASCE 7-16, which is the ratio of 85% of base shear from static analysis and base shear from the dynamic analysis.

$$V_{RSA} = \frac{SF \times I}{R}\sqrt{V_1^2 + V_2^2 + V_3^2 ...} \quad (3)$$

$$M_{RSA} = \frac{SF \times I}{R}\sqrt{M_1^2 + M_2^2 + M_3^2 ...} \quad (4)$$

$$\delta_{\mathrm{RSA}} = \frac{C_d}{R}\sqrt{\delta_1^2 + \delta_2^2 + \delta_3^2 ...} \quad (5)$$

where $V_{RSA}$ and $M_{RSA}$ are the design elastic demand of shear and moment, respectively; $\delta_{RSA}$ is the total amplified displacement or drift.

## 5. Comparison of RSA and NLRHA results

In this section, a comparison between linear RSA and NLRHA is presented. Three analytical models are used in this section which are: RSA (shown in section 4), LRSA (without $R$ and $C_d$), and NLRHA. The results of NLRHA are presented as the mean value of the peak results obtained from all the ground motions. Normalized results are presented, where floor displacement is normalized by total building height $H$, inter-story drift is normalized by story height $h$, story shear is normalized by the effective seismic weight $W$, and the overturning moment is normalized by $WH$, as shown in figure 6. Note that the title of each figure shows the building name and earthquake direction.

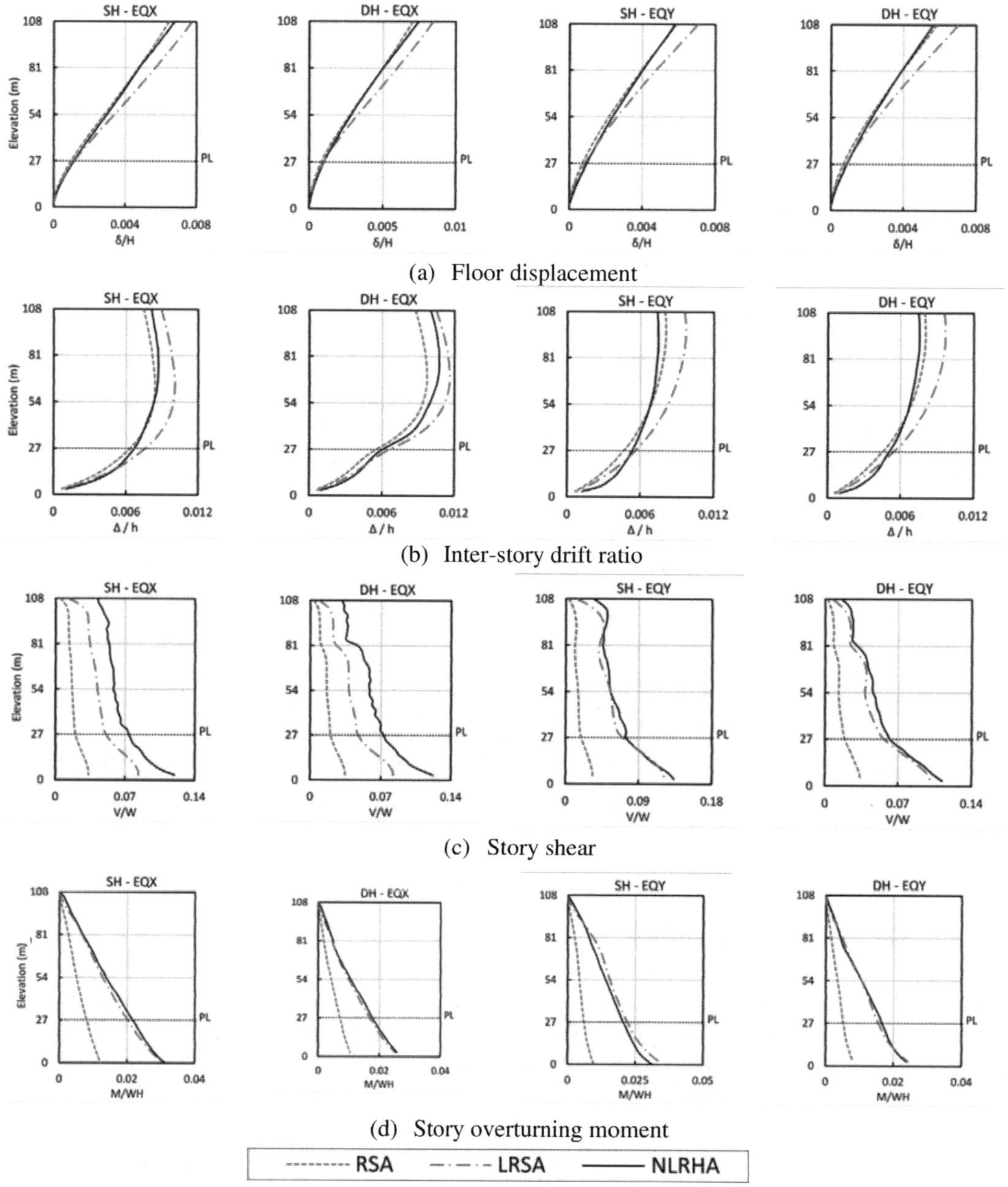


**Figure 6**. RSA and NLRHA results of (a) Floor displacement (b) Inter-story drift ratio (c) Story shear (d) Story overturning moment, for Chiang Mai buildings, due to earthquake in both directions.

It is found that the RSA procedure underestimates the force demands when compared to the actual demands obtained from NLRHA for both buildings, while it provides good estimates of floor displacements and drifts ratios for both buildings. RSA results represent seismic demands that engineers currently use in design and it is not adequate to avoid brittle shear failure, as shown in Figure 6 (c).

## 6. Inelasticity of modal responses

The modified response spectrum analysis based on higher-modes inelastic approach ($MRSA_{HI}$) proposed by Khy et al. [8] needs to determine the force response reduction factor of each mode ($R_i$) using modal pushover analysis (MPA) [16]. In this study, the inelasticity of the first three translational modes in each direction are investigated using MPA. For estimating the ($R_i$) factor, MPA was conducted using linear and nonlinear structural models. The linear model used has the same initial stiffness as the nonlinear model. The lateral load distribution for conducting MPA was applied in proportion to the distribution of mass in the plane of each floor diaphragm, multiplied by mode shape in the direction under consideration ($\mathbf{m}\phi_n$). The target roof displacement for the first translational mode in each direction was computed using the displacement coefficient method in ASCE 41-13 [14]. The target roof displacement for higher modes were assumed to be equal to the elastic response computed by modal analysis of the linear structural model considering a cracked cross-section of structural members. The base shear-roof displacement relation for the 1st translational mode in X-direction is shown in figure 7, where base shear is normalized by building weight ($W$), and roof displacement is normalized by building height ($H$). The force response reduction factor ($R_i$) is taken as the ratio of base shear of the linear to the nonlinear models at the target roof displacement. Summary of the force response reduction factors is shown in Table 4. The gravity loads of all dead loads plus 25% of live load were applied before pushover analysis. Note that the black dots in Figure 7 indicate the target roof displacement. It was found that the level of inelasticity of response in different modes was different, and the force response reduction factor ($R_i$) ranges between 1 to 2.3, which is about 2.5 to 6 times lowers than the response modification factor ($R$=6) used in the design of these buildings.

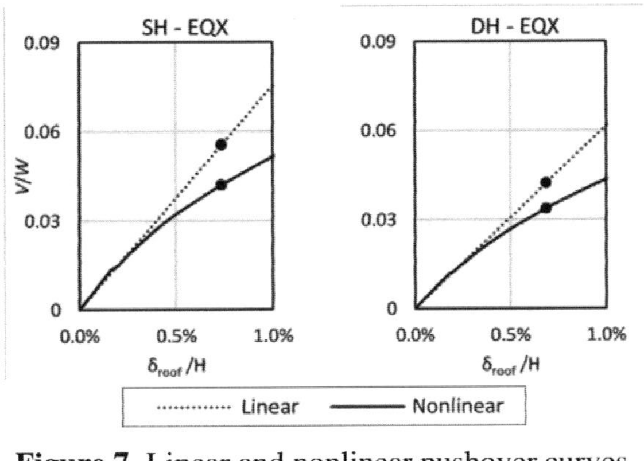


**Figure 7**. Linear and nonlinear pushover curves along with the target roof displacements of the 1st translational modes in X-direction.

**Table 4.** Force response reduction factor ($R_i$) of the first-three translational modes.

| Building | SH | | DH | |
|---|---|---|---|---|
| Direction | X | Y | X | Y |
| Mode 1 | 1.32 | 2.08 | 1.26 | 2.01 |
| Mode 2 | 1.22 | 2.37 | 1.16 | 2.28 |
| Mode 3 | 1.43 | 2.36 | 1.27 | 2.01 |

## 7. Modified response spectrum analysis

In this section, the modified response spectrum analysis (MRSA) proposed by Khy et al. [8] was conducted. Two versions of MRSA method were proposed to compute the shear forces in RC shear walls, where the first is based on higher-mode elastic approach denoted as "$MRSA_{HE}$", and the second is based on a higher-mode inelastic approach denoted as "$MRSA_{HI}$". The shear demand of $MRSA_{HE}$ and $MRSA_{HI}$ were computed from equations (1) and (2), respectively. Moreover, Khy et al. [8] suggested computing design bending moments using the RSA procedure as described in ASCE 7-16 (as shown in equation (4)). Both methods were applied using a linear model with effective stiffness corresponding to cracked cross-section properties of structural members. The total combined elastic shear demand was obtained by considering 30 modes of vibration using ETABS software. For $MRSA_{HE}$, the elastic first-mode shear multiplier is summarized in Table 5. Story shears for both buildings are shown in Figure 8.

**Table 5**. Elastic first-mode shear multiplier for $MRSA_{HE}$.

| Building | SH | | DH | |
|---|---|---|---|---|
| Direction | X | Y | X | Y |
| $(SF \times \Omega_0/R)$ | 0.80 | 0.55 | 0.82 | 0.66 |

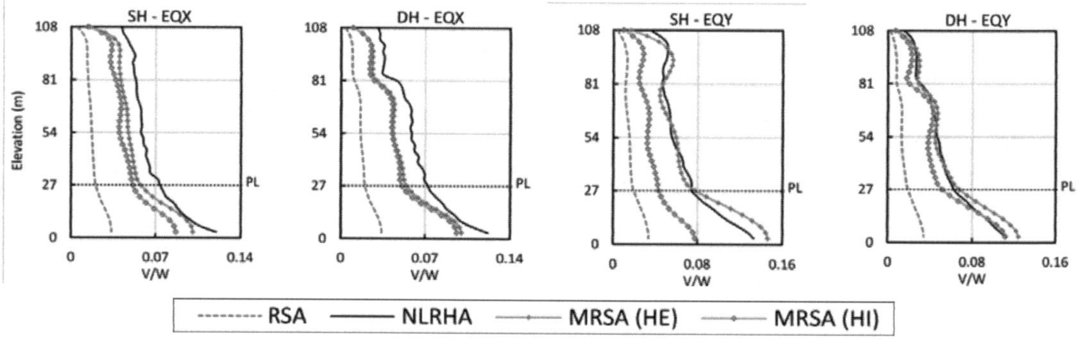


**Figure 8**. Story shear for both buildings due to earthquake in X- and Y-direction.

From Figure 8, it is found that MRSA can significantly enhance the underestimation of the RSA procedure in computing shear demands. Both versions of MRSA method provide fair results of shear demands due to seismic excitation in the X-direction. For earthquake in the Y-direction, $MRSA_{HE}$ provides accurate results of story shear, while $MRSA_{HI}$ underestimates the shear demands for SH building and provides accurate estimation for DH building.

## 8. Conclusions

The main findings of this study are summarized as follows:

1. RSA procedure underestimates force demands when compared to NLRHA results. The design (reduced) shear demand obtained from RSA procedure is not sufficient and it could lead to brittle shear failure in RC shear walls, if used in the design.
2. RSA procedure provides good estimate of floor displacements, inter-story drift ratios.
3. The inelastic behaviors of each mode were not identical, and higher modes were significantly affected by inelasticity as the first mode.
4. The $MRSA_{HE}$ method provides good estimates of shear forces for both studied RC shear wall tall buildings. The $MRSA_{HE}$ is preferred to be used in the design practice as it can be applied directly using a linear structural model.
5. The $MRSA_{HI}$ provides acceptable results for both buildings, but it underestimates shear forces due to earthquake in Y-direction for SH building.

## 9. References

[1] American society of Civil Engineers, *Minimum Design Loads and Associated Criteria for Buildings and Other Structures*. ASCE/SEI 7-16. Reston, Virginia, 2016.

[2] R. W. G. Blakeley, R. C. Cooney, and L. M. Megget, "Seismic shear loading at flexural capacity in cantilever wall structures," *Bulletin of New Zealand National Society Earthquake Engineering,* vol. 8, pp. 278-290, 1975.

[3] F. A. Najam and P. Warnitchai, "A modified response spectrum analysis procedure to determine nonlinear seismic demands of high - rise buildings with shear walls," *The Structural Design of Tall and Special Buildings,* vol. 27, p. 1409, 2018.

[4] K. Rejec, T. Isakovic, and M. Fischinger, "Seismic shear force magnification in RC cantilever structural walls, designed according to Eurocode 8," *Bulletin of Earthquake Engineering,* vol. 10, pp. 567-586, 2012.

[5] A. Munir and P. Warnitchai, "The cause of unproportionately large higher mode contributions in the inelastic seismic responses of high-rise core-wall buildings," *Earthquake Engineering & Structural Dynamics,* vol. 41, pp. 2195-2214, 2012.

[6] M. J. N. Priestley, "Does capacity design do the job? An examination of higher mode effects in cantilever walls," *Bulletin of the New Zealand Society for Earthquake Engineering,* vol. 36, pp. 276-292, 2003.

[7] K. Khy and C. Chintanapakdee, "Evaluation of seismic shear demands of RC core walls in Thailand determined by RSA procedure," *Engineering Journal,* vol. 21, pp. 151-172, 2017.

[8] K. Khy, C. Chintanapakdee, P. Warnitchai, and A. C. Wijeyewickrema, "Modified response spectrum analysis to compute shear force in tall RC shear wall buildings," *Engineering Structures,* vol. 180, pp. 295-309, 2019.

[9] Computers and Structures, *ETABS, Integrated Building Design Software, Version 18.0.1.* Berkeley, California, 2018.

[10] American Concrete Institute, *Building Code Requirements for Structural Concrete (ACI 318M-14) and Commentary (ACI 318RM-14).* Farmington Hills, USA, 2014.

[11] Computers and Structures, *PERFORM-3D, Nonlinear Analysis and Performance Assessment of 3D Structures, Version 7.0.0.* Berkeley, California, 2018.

[12] Bangkok Metropolitan Administration. (2001). *Bangkok Building Control Law, B.E. 2544.* Available: http://203.155.220.230/bmainfo/law/law.php?t=031

[13] J. B. Mander, M. J. Priestley, and R. Park, "Theoretical stress-strain model for confined concrete," *Journal of Structural Engineering, ASCE,* vol. 114, pp. 1804-1826, 1988.

[14] American Society of Civil Engineers, *Seismic Evaluation and Retrofit of Existing Buildings.* ASCE/SEI 41-13. Reston, Virginia, 2013.

[15] K. Khy and C. Chintanapakdee, "Improved response spectrum analysis procedure for design of reinforced concrete tall buildings," Chulalongkorn University, Bangkok, 2018.

[16] A. K. Chopra and R. K. Goel, "A modal pushover analysis procedure for estimating seismic demands for buildings," *Earthquake Engineering and Structural Dynamics,* vol. 31, pp. 561-582, 2002.

IOP Conf. Series: Materials Science and Engineering 1044 (2021) 012006 doi:10.1088/1757-899X/1044/1/012006

# Automated Shape Adjustment of Interlocking Joints for Structurally Informed Design of Masonry Block Assemblages

**Elham Mousavian*, Claudia Casapulla**

Department of Structures for Engineering and Architecture, University of Napoli Federico II, Via Forno vecchio, 36, 80134 - Napoli, Italy

*elham.mousavian@unina.it

**Abstract.** This paper introduces a tool to design structurally informed assemblages of interlocking blocks. First, the static problem of limit analysis is extended to assemblages with openings and a running bond pattern. Then, a method to measure the assemblage infeasibility due to the sliding resistance violation (namely sliding infeasibility) is described. An optimization procedure is finally presented to minimise the measured sliding infeasibility by automatically changing the shape of interlocking joints.

## 1. Introduction

The mechanical improvement of block assemblages through finding the best interlocking joint shapes has recently been of significant interest to the fields of architecture and structural engineering. Several studies have addressed the relation between different joint shapes and the assembly's structural performances [1-3], though mostly investigating the influence of limited geometric parameters of modular interlocking blocks on the wall in-plane or out-of-plane behaviours. Few methods have been developed to design structurally wise shapes for dry interlocking joints: Yao et al. [4] have developed a tool that first provides the user with an interface to design the desired joints between the rigid parts, then analyse the stability of the model and suggest how to remove the instability; Rippmann [5] has shown that the bearing capacity of a compression-only, rigid, and frictionless body does not change after being partitioned by non-planar (interlocking) joints normal to the flow of axial forces. The authors of this paper have developed a tool within the European SiDMACIB Project, No. 791235 MCSA_IF, as one of the first efforts to design structurally sound masonry assemblages of interlocking blocks with finite shear and friction. The SiDMACIB tool is based on a two-step procedure: it first extends the static problem of limit analysis to orthotropic corrugated joints [6], and then quantifies the structural infeasibility due to the violation of the sliding resistance, namely "sliding infeasibility measure" (SIM) [7]. The standard and non-standard limit analysis [8-10] is a powerful method to analyse masonry structures with more computational efficiency when compared to other well-known methods including finite element [11] and discrete element [12] analyses. On the other hand, the concept of measuring how far a model is from being structurally feasible has firstly been introduced by Whiting et al. [13] and it is a powerful tool to minimize the structural infeasibility by automatically or manually adjusting the initial shape or material parameters assigned by the user. This paper presents an optimization method which automatically tunes the interlocking joint shapes to minimize the SIM, i.e., to remove the structural infeasibility of the initial model. The optimization particularly changes the orientation of the locks at the interlocking joints to minimize the sliding infeasibility.

## 2. Equilibrium problem of an assemblage of interlocking blocks with the running bond pattern

A plugin for the modelling and analysis of single layer assemblages with a stack bond (aligned joints) pattern was recently developed and presented in [6], where the static problem of limit analysis was extended to orthotropic interlocking joints. In this paper, the algorithms of this plugin are updated efficiently to model single layer shells with the running bond pattern and the openings inside the shell. To this aim, a shell with either stack or running bond is structurally modelled as an arrangement of the units with aligned joints. While a unit of the stack bond model shows a single real block, each unit of a running bond shell represents half of a real block (figure 1-left). All the joints shared between the units of a stack bond shell are the peripheral joints *pj* around the real blocks. On the contrary, there are two types of joints between the units of a running bond shell: 1) peripheral joint *pj* between two real blocks and 2) rigid joint *rj* between two units of a real block in the *U* direction, which are in fact part of the rigid block. The plugin categorises the block faces in four sets of *uv*0, *uv*1, *uw* and *vw* faces, which are on the shell intrados and extrados, and intermediate faces in the *U* and *V* directions. Given a lock orientation and number for each set, the *pj*s of the set can be formed as interlocking joints.

The external forces are applied to the unit centroids, i.e., for a running bond shell two external forces are applied to two units of one block. The internal forces on the *pj*s and *rj*s equilibrate the applied external forces while satisfying the joint normal and shear constraints (figure 1-right). A *rj* is part of a rigid block and therefore has infinite normal and sliding resistances, i.e., it can take any value; on the contrary, a *pj* is tensionless and has finite sliding resistances. The joints between the main body of an interlocking block and its locks, which are considered rigid, can fail; it is assumed that a lock can be cracked and separated from the block main body due to lock bending or torsion-shear failures. Two blocks can also be displaced with respect to each other along the locks, when the joint frictional resistance is not enough. Adopting the Livesley's concave model [9] a *rj* is abstracted to four contact points at the joint corners and a *pj* is abstracted to a number of contact points distributed on the lock centre lines C.L. Three internal forces are considered at each contact point of either *pj* or *rj*, one normal $r^{n}$ and two tangential $r^{t1}$ and $r^{t2}$ to the joint. For a contact point on a *pj*, $r^{pj,n}$ is constrained to be in compression, $r^{pj,t1}$ is parallel to the C.L. and must satisfy the frictional resistance, and $r^{pj,t2}$ is normal to the C.L. and should be lower than 0.33 of the total lock shear resistance $T_0 = \tau_k\, s\, b$ when three contact points are assumed distributed on each C.L. [6], being $\tau_k$ the material shear strength, and $s$ and $b$ the lock width and length, respectively. The openings can be modelled through the removal of selected units and setting the external forces applied to the removed units to be zero, while the joints shared between the removed units are excluded from the joints at which the internal forces can exist. Finally, the constrained equilibrium equation can be written as:

$$\begin{cases} C_{eq,\theta}.\vec{r} + \vec{E} = 0 & & \text{Equilibrium equation} \\ \text{subjected to:} & \overrightarrow{r^{pj,n}} \le 0 & \text{compression constraint} \\ & \left|\overrightarrow{r^{pj,t1}}\right| \le \mu \left|\overrightarrow{r^{pj,n}}\right| & \text{friction constraint} \\ & \left|\overrightarrow{r^{pj,t2}}\right| \le 0.33\, T_0 & \text{shear constraint} \end{cases} \quad (1)$$

where $C_{eq,\theta}$ is the equilibrium coefficient matrix for a generic set of lock orientations $\theta = \{\theta_{uv0}, \theta_{uv1}, \theta_{uw}, \theta_{vw}\}$. Changing $\theta$, $C_{eq,\theta}$ and the structural feasibility change.

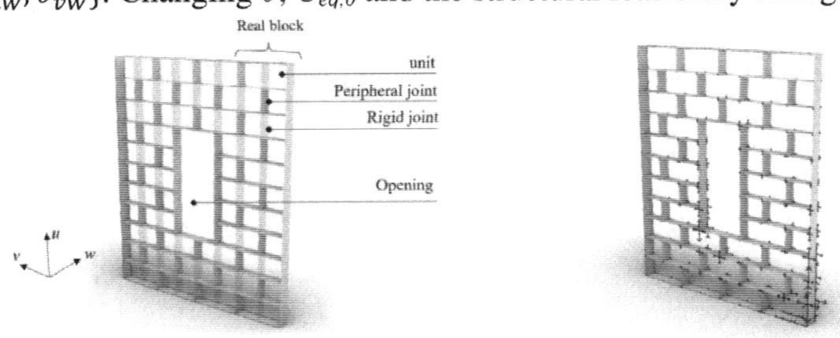


**Figure 1.** Left: a wall with one opening and running bond. Right: Internal force at the limit state

*2.1. validation*

The static problem has been implemented to develop a grasshopper plugin, using C#, while the MATLAB's lsqlin has been used to solve the linear equilibrium equation. The benchmark problem chosen for validating the implemented static problem is a vertical wall with the running bond pattern and one opening, subjected to in-plane loading. The chosen wall is in fact the third benchmark wall problem provided by Gilbert et al. [10] (figure 1). It is composed of blocks with 4 × 1.75 units of length subjected to one unit of vertical force (block weight) and one unit of lateral force. Also, the friction coefficient is considered 0.65. The collapse load factors λ for associative friction reported in [10] and this paper are 0.40369 and 0.40368, respectively, and this good matching can validate the implemented method.

## 3. SIM optimization

This section briefly introduces the method to quantify the sliding infeasibility of interlocking assemblages (sliding infeasibility measure SIM). The method was elaborated in detail in [7]. To measure the sliding infeasibility, the sliding constraints can be removed from Eqation (1) so that the internal forces can take any value. For each contact point, the absolute value of the tangential force violating the sliding valid ranges can then be considered as the local SIM. Summing (functions of) the local SIMs at all contact points, the overall SIM of the assemblage can be calculated. For an infeasible assemblage, equation (1) with the removed sliding constraints may find several solutions, i.e., several SIM might exist. To find the solution with the lowest SIM, equation (1) can be reformulated as follows. First, the tangential force at each contact point $r^{t}$ is decomposed into two new variables $r^{ta}$ and $r^{tb}$, where $r^{ta} + r^{tb} = r^{t}$; while $r^{ta}$ has to satisfy the sliding constraints, $r^{tb}$ can take any value and therefore $r^{t}$ can also take any value. Then, an optimization problem is formulated which tries to minimize $r^{tb}$, making $r^{t}$ as close as possible to $r^{ta}$ (within the sliding valid range), while the assemblage is constrained as follows:

$$\begin{cases} SIM_{\theta} = \min \sum_{\eta=1}^{L} [(r_{\eta}^{pj,t1b})^2 + (r_{\eta}^{pj,t2b})^2] & \textit{Objective function} \\ \textit{subjected to:} \quad C_{eq,\theta}.\vec{r} + \vec{E} = 0 & \textit{Equilibrium equation} \\ \overrightarrow{r^{pj,n}} \leq 0 & \textit{compression constraint} \\ \left|\overrightarrow{r^{pj,t1a}}\right| \leq \mu \left|\overrightarrow{r^{pj,n}}\right| & \textit{friction constraint} \\ \left|\overrightarrow{r^{pj,t2a}}\right| \leq 0.33\, T_0 & \textit{shear constraint} \end{cases} \quad (2)$$

where $r_{\eta}^{pj,t1b}$ and $r_{\eta}^{pj,t2b}$ are the tangential forces parallel and normal to the lock at the generic contact point η, respectively, $L$ is the total number of contact points, $\vec{r}$ is the vector of variables containing the internal forces $r^{t1a}$, $r^{t1b}$, $r^{t2a}$, $r^{t2b}$, and $r^{n}$ at all the contact points. The designer can adjust $\theta$ manually to minimize the SIM and remove the sliding infeasibility of an assemblage of interlocking blocks, which is in fact the result of the objective function at the optimal solution.

## 4. Lock-orientation optimization

This section introduces the lock-orientation optimization algorithm by which the initial lock orientations assigned by the user are automatically tuned to minimize the SIM. Before elaborating this optimization, first the concept of function in programming is introduced. In computer programming, a function is a black box containing a block of codes getting a number of inputs, executing several tasks and returning outputs. This black box can then be called again so that its output is a function of the selected inputs. To develop the lock-orientation optimization, a function is defined to compute the SIM and its name is simply "SIA" standing for sliding feasibility analyser. The SIA inputs are a set of lock orientations $\theta = \{\theta_{uv0}, \theta_{uv1}, \theta_{uw}, \theta_{vw}\}$ and another set, called "assigned_inputs", including all the parameters assigned by the designer, such as the lock numbers ($N = \{N_{uv0}, N_{uv1}, N_{uw}, N_{vw}\}$), coordinates of the block corners, external loads and moments, indices of the block faces acting as supports, and material properties μ and $\tau_k$. When a lock number in set $N$ is assigned zero, the related faces are considered to be flat and therefore, the related lock orientation does not change the SIM value. Using MATLAB, SIA can be written as shown in figure 2.

```
function SIM = SIA (theta, assigned_inputs)

%% C_ineq,RHS_ineq,C_eq,E construction%%...%%
%% C_obj construction%%...%%
%% r0 construction%%...%%
%% Optimization
    f = @(r) C_obj* r.^2; %objective function
    [r,SIM] = fmincon(f,r0,C_ineq,RHS_ineq,C_eq,E); %optimization
end
```

**Figure 2.** MATLAB code snippet for SIA function

This function first constructs the matrices and vectors of equation (2), including: the coefficient matrix "`C_eq`" and the right-hand side vector "`E`" of the equality constraint (equilibrium equation in equation (2)); the coefficient matrix "`C_ineq`" and right-hand side vector "`RHS_ineq`" of the inequality constraint (normal and sliding constraints of equation (2)); the coefficient vector "`C_obj`" multiplied by optimization variables "`r`" to the power 2 to define the equation (2) objective function; and the initial value of the variables "`r0`". How they are constructed is folded (hidden) in the code snippet. Then, the optimization objective function "`f`", whose variables are the internal forces "`r`", is defined using the syntax "`@r`" and the MATLAB's fmincon solves this quadratic function. Finally, the SIM that is the value of the `f` at the optimal solution is returned as the SIA output and the SIM can be called again as a function of $\theta$: `SIM = @(theta) SIA (theta, assigned_inputs)`. The next step is to define another black box function which gets the inputs assigned by the user in a grasshopper component (through C#) and returns the optimal $\theta$. It is named LOO, standing for the lock-orientation optimizer (figure 3).

```
function theta = LOO (assigned_inputs)

SIM = @(theta) SIA (theta, assigned_inputs);        %objective function

theta0 = [0,0,0,0];          %initial theta
lb(1:4)=0;                   %theta lower bound
ub(1:4)=pi;                  %theta upper bound

theta = fmincon(SIM,theta0,[],[],[],[],lb,ub);   %optimization
end
```

**Figure 3.** MATLAB code snippet for LOO function

This function sets up all the necessary materials for the lock orientation optimization, i.e.: a set of initial values for $\theta$ from which the optimization starts (`theta0`); the objective function `SIM` whose variable is `theta`; and the upper "`ub`" and lower "`lb`" bounds for $\theta$ which are zero and $\pi$. LOO inputs are the "`assigned_inputs`" introduced above, including all the shape and material properties needed to find the SIM. LOO gets the `assigned_inputs` from the grasshopper component and passes on them to the SIA function that is invoked in LOO. MATLAB's fmincon is used to solve the shape optimization. Finally, the optimal $\theta$ (i.e., the optimal $\theta_{uv0}$, $\theta_{uv1}$, $\theta_{uw}$, and $\theta_{vw}$) is returned as the LOO output.

### *4.1 Validation*

To validate the shape optimization, a simple setup of two blocks is considered, where the lower block is fully fixed and the upper one is subjected to a lateral force. This setup was previously addressed in [6] to study the torsion-shear resistance of interlocking blocks. The setup dimensions and the forces applied to the block centroids are shown in figure 4. The shear strength and the friction coefficient are 0.25 $N/m^2$ and 0, respectively. In this model, only the *vw* faces are considered to be interlocking. The lateral force imposed to the upper block is the collapse load for the case in which the two considered locks are normal to the force.

First, the SIM is obtained manually by solving equation (2) for different values of $\theta_{vw}$ changed by the user. As shown in figure 4, when the locks are parallel to the lateral force the system is infeasible since the frictional resistance is zero and cannot avoid sliding at the interface. The largest SIM is recorded for this case (7.45 $N^2$). Changing $\theta_{vw}$, the sliding resistance is increased since both joint shear and frictional resistances are involved in the sliding prevention. For $\theta_{vw} = 1.57$ rad, the SIM is $2.2 \times 10^{-4}$, meaning that the sliding infeasibility is removed when the locks are normal to the lateral force. In the second step, the lock-orientation optimization is executed to the same setup, setting the lock orientation initial values as $\theta^0 = \{0, 0, 0, 0\}$, i.e., the optimization starts from the case in which the two locks are parallel to the lateral force, causing the largest SIM. The assigned lock numbers are $N = \{0, 0, 0, 2\}$, which means that only the *vw* faces are interlocking with two locks. As a result, only the initial value of $\theta_{vw}$ will be changed during the optimization. The optimal $\theta_{vw}$ found by the shape optimization is 1.566477 rad and its corresponding SIM is $3.2 \times 10^{-4}$ which is very close to the minimum SIM found manually ($2.2 \times 10^{-4}$).

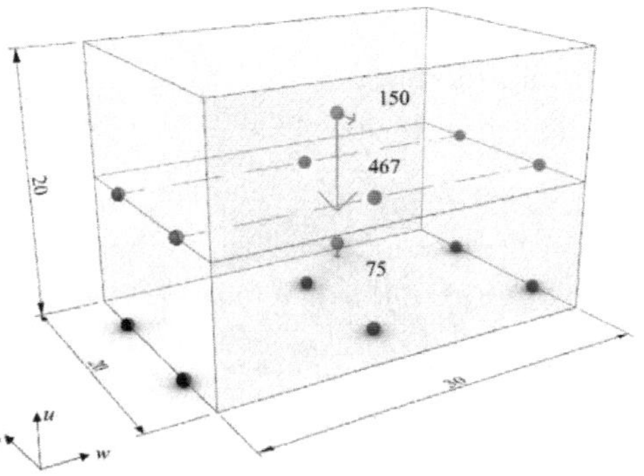


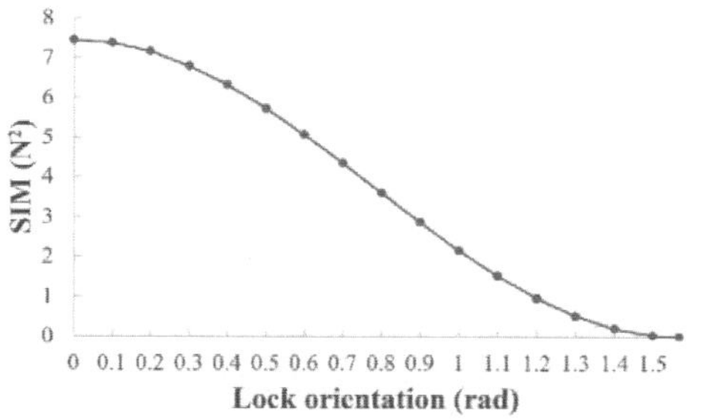


**Figure 4.** Left: The validation setup (the black points are the supports). Right: Relation between the SIM and lock orientation found manually

## 5. Conclusions

This paper demonstrated and validated an extension of the static problem of limit analysis to assemblages of interlocking blocks with the running bond pattern and openings. A method was proposed to measure the assemblage sliding infeasibility and a shape optimization was developed to automatically adjust the lock orientations to minimise the measured infeasibility. The optimization was validated by comparing the SIMs for different lock orientations obtained manually and the minimum SIM automatically found by the optimization.

## Acknowledgments

This project has received funding from the European Union's Horizon 2020 research and innovation programme under the Marie Sklodowska-Curie Grant Agreement No. 791235.

## 6. References

[1] Liu H, Liu P, Lin K and Zhao S 2016 Cyclic behavior of mortarless brick joints with different interlocking shapes *Materials* **9(3)** 166

[2] Dyskin A V, Estrin Y and Pasternak E 2019 Topological Interlocking Materials *Architectured Materials in Nature and Engineering* ed Y Estrin (Berlin: Springer) pp 23-49

[3] Javan A R, Seifi H, Lin X and Xie Y M 2020 Mechanical behaviour of composite structures made of topologically interlocking concrete bricks with soft interfaces *Materials & Design* **186** 108347

[4] Yao J, Kaufman D M, Gingold Y and Agrawala M 2017 Interactive design and stability analysis of decorative joinery for furniture *ACM Transactions on Graphics* (TOG) **36(2)** 20

[5] Rippmann M 2016 *Funicular Shell Design: Geometric approaches to form finding and fabrication of discrete funicular structures* (ETH Zurich: Doctoral dissertation)

[6] Mousavian E and Casapulla C 2020 Structurally informed design of interlocking block assemblages using limit analysis *Journal of Computational Design and Engineering* **7(4)** 448-68

[7] Mousavian E and Casapulla C 2020 Quantifiable feasibility check of masonry assemblages composed of interlocking blocks *Advances in Engineering Software* **149** 102898

[8] Heyman J 1995 *The stone skeleton: structural engineering of masonry architecture* (Cambridge: Cambridge University Press)

[9] Livesley R K 1992 A computational model for the limit analysis of three-dimensional masonry structures *Meccanica* **27(3)** 161-72

[10] Gilbert M, Casapulla C and Ahmed H M 2006 Limit analysis of masonry block structures with non-associative frictional joints using linear programming *Computers & Structures* **84(13-14)** 873-87

[11] Lourenço P B, Rots J G, and Blaauwendraad J 1995 Two approaches for the analysis of masonry structures: micro and macro-modeling *Heron* **40(4)** 1-28

[12] Bagi K and Orosz Á 2020). A new variable for characterising irregular element geometries in experiments and DEM simulations *Proc. Communications of the ECMS* vol 34(1) (United Kingdome: European Council for Modeling and Simulation) pp 256-260

[13] Whiting E, Ochsendorf J and Durand F 2009 Procedural modeling of structurally-sound masonry buildings *ACM Transactions on Graphics (TOG)* **28(5)** 112

IOP Conf. Series: Materials Science and Engineering 1044 (2021) 012007 doi:10.1088/1757-899X/1044/1/012007

# Innovation Architecture: Pickle Processing Industry "La Pukarina" – Province of Huancayo, Peru

**Katherine Toralva Garay[1*], Jorge Sihuay Maraví [1*], Nabilt Moggiano[2*]**

[1]Department of Architecture, Universidad Continental, Huancayo, Peru
[2]Engineering Research Unit, Faculty of Engineering, Universidad Continental, Huancayo, Peru

Email: ktoralva@continental.edu.pe

**Abstract.** The Pucara area has land with horticultural potential, so the project proposes the construction of "La Pukarina" Pickle Processing Plant. The proposal is the result of the diagnosis of horticultural production in the area to give added value to your vegetables; producing pickles, promoting associative business work. Using the Location Factors Rubric, made up of 9 indicators, the land for the Processing Plant was located, with an area of 24 823.25 m$^2$, accumulating 34 points. The design of the plant is born from a formal search, using the geometry of the medium as the corridors and ceilings. In the same way, the Mahoney Table will be seen, a bioclimatic design method for its distribution. After the economic analysis of the project, the initial investment of 445282.73 USD was established in the first year, financed at a bank interest of 7%, and financed in 6 years. The market study requires the target population, made up of men and women between 25 and 39 years old, of whom 58% consume pickles every two weeks, especially an organic vegetable base in the area.
Finally, the Pickles Processing Industry "La Pukarina" is proposed as a technically and economically viable alternative to up 100.8 %.

## 1 Introduction

The Pucara area, located in the province of Huancayo (Peru) enjoys land with high potential for growing vegetables, its population is dedicated to intensive agriculture of horticultural products well known in central Peru and part of Lima, especially those produced in the populated center of Raquina, Pucapuquio, Asca [1], all produced under irrigation because they are immersed in the Chanchas river micro-basin, whose harvests are marketed in nearby markets and apart for self-consumption, highlighting: Chinese onion, radish, beet, carrot, spinach, among others [2]. After diagnostic work on horticultural production in the Pucara area, the Project: Innovation Architecture for the production of the Pickles Processing Industry "La Pukarina" was proposed.

The project seeks to propose a favorable infrastructure, thus ensuring markets and profitability of its production, knowing that, for the most part, the horticulturists in the area are small producers, smallholders; that they practice traditional agriculture with low technological capacity, without financing, with yields that can be surpassed giving an added value to their vegetables; producing pickles and fundamentally with associative business work.

The work addresses the design of a food canning and production plant, specifically the production of Pickles, located on "5 de Febrero Avenue" in the Pucara district, Huancayo Province, Junin Region; in an area of 24 823.25 m$^2$ and a perimeter of 694.53 linear meters, zoned as District Commerce which is compatible with light industry. No major geological structures were found in the area, making it a

IOP Conf. Series: Materials Science and Engineering 1044 (2021) 012007 doi:10.1088/1757-899X/1044/1/012007

safe area. The soil for agriculture is composed of conglomerates of clay, silt and sand, the place to build the plant has gravels which allows the processing plant to be built safely.
The turnip, beets, and carrots will be processed, in their first stage as they are the most produced vegetables in the area accompanied by spices, aromatic herbs, condiments, vinegar, etc. that is to say preserved by means of natural fermentation.

The process begins with reception machinery such as scales; production machinery that highlights the spray washer, conveyor belts, autoclave, cutter, filler, vinegar tank, government liquid dispenser, etc. Expedition machinery such as labeler, pasteurizer, packaging, etc. Warehouse area with a forklift and cleaning area. An orchard is included, for the introduction and subsequent technical assistance to the farmers of the area, for the treatment of wastewater a biodigester is included in order to treat these waters and reuse them in the area dedicated to plantations and obtaining organic fertilizer. The infrastructure is divided into 4 zones: Administrative, Commercial, Production, and Services.

## 2 Method

The present study is of a descriptive type since it collects theoretical, documentary, and interview information considering the characteristics of its components: the vegetable processing industry and the innovative architecture proposed in the study. To collect information, the fieldwork was carried out, visiting the communal authorities of Pucará and Raquina in order to request authorization to locate the plant of proccessing.

The Pucara district is located at 3,340 m. a. s. l. with an area of 110.49 $km^2$, 20 km from the city of Huancayo, with a population of 5,063 inhabitants according to the last national census by INEI (2007), is divided into 10 annexes and scattered settlements: Asca, Puca Puquio, Raquina, Talhuis, Jatun Suclla, Pachachaca, Patala, Marcavalle, Chucos, Dos de Mayo and dispersed population.
The area of Raquina, an eminently horticultural town, was analyzed on the banks of the Chanchas river in its origin and with five springs that supply water to its vegetable crops, the annex has a narrow road, surrounded by mountains, with only single-phase electric energy. Population of "Anexo de Raquina" is dedicated to agriculture, especially the cultivation of organic vegetables. Vegetable production, mostly a smallholder activity, is produced in small plots that contribute to family nutrition and as a source of income for their families.

The production of vegetables is predominant crops are the turnip, onion, beet, radish, Chinese onion, among others, whose production costs and yield per hectare vary according to the species produced and the use of agricultural technology or traditional plowing, adding the use of organic fertilizers and fertilizers, leaving aside the use of chemical products because organic agriculture is promoted in the place. Yields vary between 5,000 to 12,000 kg/ha depending on the species to be cultivated with an average investment of 570 USD to 1500 USD per ha. The cultivation area for vegetables under irrigation is 864 ha between Raquina and Pucará.
To determine the location of the Pickle Processing Plant, the Location Factors Rubric was used, which considers [4] among other aspects: Proximity to raw materials, proximity to the target market, availability of labor, the supply of electrical energy, drinking water, sewage, wind orientation, access to transportation, industrial waste management, with Scale values from 1 to 5 respectively. The fundamental reason for choosing the land was due to its location close to the production centers, proximity to the highway that facilitates the transport of the processed product to the target markets.
Three lands suitable for the construction of the Processing Plant were taken into account. Proposal N° 1: is found in Jr. Cabo Saavedra, Pucara district, Huancayo province; with an area of 3 169.29 $m^2$. Proposal N° 2: is found on Av. Catalina Huanca, Pucara district, Huancayo province; with an area of 3 086.0 $m^2$. Proposal N° 3: is located on "5 de Febrero Avenue", Pucara district, Huancayo province; with an area of 24,823.25 $m^2$.

To find trends to industrialize vegetables, a significant sample was applied to 100 respondents, applying the theory of statistical sampling based on the Economically Active Population of Junin considered in the 2017 census of 538.286 inhabitants [3].

*2.1Productive strategy*

Pickles are vegetables preserved in vinegar, they are important because they keep vegetables longer, which must be kept in hermetically closed and previously sterilized glass containers; whose production is carried out throughout the year, in our proposal it is carried out in 8-hour days a day, respecting current legal provisions of the Ministry of Labor, having as working hours: Entering at 08:00 until 17:00 with an hour that includes rest, snacks and maintenance of machinery.

During the snack, the cleaning staff will carry out the cleaning of the production facilities including offices, warehouses, sanitary services, among other environments.

240 days a year will be worked, that is to say with 20 days a month; the week will consist of 5 days of productive activity and the Saturdays of each week will be used for cleaning and maintenance of the production machinery.

*2.2 Productive program*

The production process begins with the reception and quality control of the vegetables, then a heat treatment is given, the vegetables are minced in different ways: in pieces, halves, julienne, etc. They are packed and finished with their storage and later commercialized in the markets of the Mantaro Valley.

## 3 Results

Land Location

Applied the Rubric, the Proposal N° 3 was selected, where the land is located on "5 de Febrero avenue", Pucara district, with area of 24 823.25$^2$ accumulating 34 points as shown in table I.

**Table 1.** Indicators to determine the location of the pickle processing industry

| **Proposal** | **Proposal N° 1** | | | | | **Proposal N°2** | | | | | **Proposal N°3** | | | | |
|---|---|---|---|---|---|---|---|---|---|---|---|---|---|---|---|
| **Scale** | 1 | 2 | 3 | 4 | 5 | 1 | 2 | 3 | 4 | 5 | 1 | 2 | 3 | 4 | 5 |
| **Indicators** | | | | | | | | | | | | | | | |
| **Proximity of raw materials** | | | | | X | | | | X | | | | | | X |
| **Close to the target market** | | | | X | | | | | X | | | | | X | |
| **Availability of labor** | | | X | | | | | X | | | | | | X | |
| **Electric power supply** | | X | | | | | X | | | | | | | | X |
| **Potable water supply** | | | X | | | | | X | | | | | X | | |
| **Drainage supply** | X | | | | | X | | | | | X | | | | |
| **Wind orientation** | X | | | | | | X | | | | | | | | X |
| **Access to transportation** | | | | X | | X | | | | | | | | X | |
| **Industrial waste management** | | X | | | | X | | | | | | | X | | |
| **Total score** | 25 | | | | | 20 | | | | | 34 | | | | |

[1] Poor, [2] Regular, [3] Good, [4] Very good, [5] Excellent

The first step for the market study of pickles; was to determine the target population to arrive at the market the Pickles "La Pukarina" which was made up of men and women from 25 to 39 years, the figure 1 shows the 58% consume pickles biweekly in meals with friends, especially based on vegetables organic from the area. For this purpose, a survey was prepared with 10 questions, reaching a survey of 100 people throughout the city of Huancayo, the non-probability sampling technique was used.

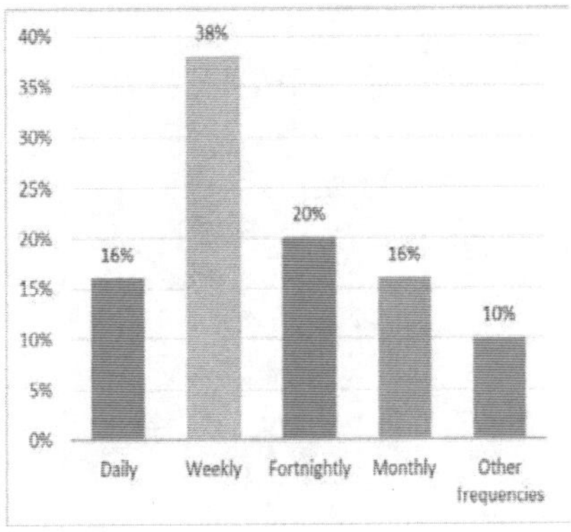


**Figure 1.** Pickles consumption frequency in the Mantaro Valley

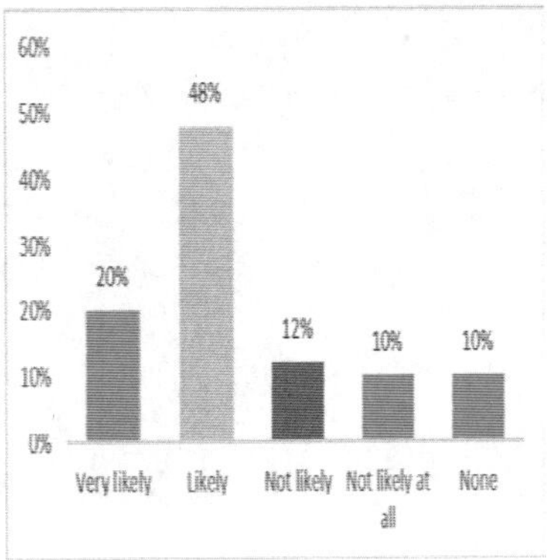


**Figure 2.** Projection of final consumer purchase of pickles

Figure 2 shows the question: Would you buy Pickles from a local company with organic products? 68% of the respondents would probably buy because their prices would be reasonable, strengthening the viability of the Project.

## 4 Architectural Proposal

The work addresses the design of a Pickle Production and Processing Plant in the Pucara area. A marginal strip of 50 m is proposed, 2 access roads, the first connect Pucara with our raw material; Raquina, the second for the entrance of workers and visitors. The green barrier for the welfare of workers against noise emission.

- The guiding idea is born from a formal search, using geometry.
- The design starts from a sociocultural dimension and the concepts of the environment.
- The formal spatial values of the area were used: patios and corridors and as envelopes, sloping roofs, favorable to the climate, and identification to the context.
- The internal spatial forms have dimensions appropriate to the production process, lighting, ventilation, views, and comfort. In the compositional part, horizontal and vertical spatial fluidity are proposed with integrations relevant to the activities.
- It begins with 3 prisms with a rectangular base, compositionally related to each other.

Assembly techniques

Two assembly techniques were used. The first one is the penetration assembly in the commercial zone with the administrative following the joint assembly for the service area.

To obtain a better condition for the Plant and the workers, the Mahoney Table [5] was used, a bioclimatic design method that evaluates climatic conditions to compare them with a comfort limit and evaluate them to reference the type of bioclimatic resource to be used [6].

The data obtained at the Viques meteorological station [7] was processed to the climatic data table for the analysis of the Mahoney indicator table. After this, it indicates the North-South orientation, the spread of compact configuration, the size of the openings between 50% - 80%, which can be seen in the Production area and the light and isolated roofs. Figure 3 shows the design of the Production Zone, the pickling process [8], [9] and [10] was considered, starting with the reception and quality control of the vegetables, then treatment is given thermal, the vegetables are chopped in different ways: in pieces, halves, julienne, etc. They are packed and finished with their storage and later commercialized in the markets of the Mantaro Valley [11]. The turnip, beet, and carrot processing will be prioritized, as it is a vegetable of greater production in the area accompanied by spices, aromatic herbs, condiments, vinegar, etc. preserved by natural fermentation [12]. The administrative area has spaces for the development of its activities: Advertising Office, in charge of promoting the products; Shipping and sales office to market pickles, an institutional hall as figure 4 shows for workers and clients, Administrative Office, Human Resources, Accounting, and Archives.

**Figure 3.** Production area of the pickle processing industry.

**Figure 4.** Institutional dining room for workers and visitors.

The commercial area has two exhibition rooms, a multipurpose room, the reception and entrance of visitors to the plant. The service area has a nursery for the introduction and subsequent technical assistance to the local horticulturists, to treat wastewater from productive activities and others; A biodigester is included in order to treat these waters and reuse them in the area dedicated to plantations. Finally, composting to obtain organic fertilizer.

## 5 Benefits and costs

The costs and benefits are made up of investment in land, construction (infrastructure), machinery, labor, raw materials; with an initial investment approximately 445282.73 USD in the first year financed at a bank interest of 7% payable in 6 years. "La Pukarina" S.A. integrated by 20 community members of the Peasant Community of Pucara - Huancayo, who will produce 1500 units of pickles daily, for the markets of the Mantaro Valley, Lima and others. From the Investment Analysis, we state that the rate of return (IRR) after investing in the project will be 115% with profitability of 100.88%. The Net Present Value (NPV) is greater than 0, which means that the capital would be recovered in the fifth year and the amount of profits will exceed the initial investment and the capital will be increased.

## 6 Conclusion

After the study, the "La Pukarina" Pickle Processing Industry is proposed as a technically and economically viable alternative 100.8 %, since in Mantaro Valley there is no logistics chain for the distribution of horticultural products, nor for processed products, which in the long term generates losses to producers in the Pucara area.

A new infrastructure and the start of a vegetable product processing industry in the region respect the environment and its surroundings, enables the construction of a Pickle Production and Processing Plant with materials from the region in the Pucara, province of Huancayo, Peru.

## 7 References

[1] M. Laporte, G. Faure, y Le Gal "Diversidad de las explotaciones agrícolas en los sistemas irrigados del valle del Mantaro y acceso de los productores al mercado". Lima: PSI 2008.
[2] Gobierno Regional de Junín "Potencial de cultivos y ganadería en la región Junín" Huancayo,2003.
[3] INEI, Censo Nacional 2017, Tomo1 Junín, 2018.
[4] L. Ordoñez, "Localización y distribución de plantas agroindustriales". Colombia: Universidad Nacional de Colombia, pp. 15-27, 2001.
[5] C. Mahoney, "Tablas Mahoney" Inglaterra: Instituto de Arquitectura Tropical, 1971.

[6] L. Machuca, J. Molina, R Espinoza. "Estudio climático de Vilcallamas arriba y análisis de indicadores bioclimáticos de aplicación potencial" - XIX Simposio Peruano de Energía Solar, 2012.
[7] Universidad Continental "Boletín meteorológico enero – abril. Huancayo" UC ,2017.
[8] M. Blanco. "Procesamiento de frutas, hortalizas y especias en pequeña escala. Alternativas" San José Costa Rica ,1992
[9] A. Bartholomai "Fábrica de alimentos: procesos, equipamientos, costos". Zaragoza, Acribia ,1991
[10] D. Colquichagua. "Encurtidos" Lima Tarea Asociación Gráfica Educativa, pp. 10-13,1998.
[11] F. Duran "Seguridad alimentaria cultivando hortalizas" Colombia Grupo Latino editores, pp. 105 – 116, 2011.
[12] G. Paltrinieri, F Figuerola. "Procesamiento de frutas y hortalizas mediante métodos artesanales y de pequeña escala". Manual Técnico. Santiago: Oficina Regional de la FAO para América Latina y el Caribe,1998.

IOP Conf. Series: Materials Science and Engineering 1044 (2021) 012008 doi:10.1088/1757-899X/1044/1/012008

# Optimizing Knitted Fiber-Reinforced Composite Carabiners

**William Fu[1,*] and Behnam Ahmadikia[2]**

[1]Neuqua Valley High School, 2360 95th St., Naperville, IL 60564, USA
[2]Department of Mechanical Engineering, University of California, Santa Barbara, CA 93106, USA

[*]Corresponding Author: william.fu2288@gmail.com

**Abstract.** Due to the looping geometry of the knitted fibers, composites reinforced by knitted fabrics excel particularly in their light weight, high impact strength, and adaptable contourability for complex shapes. Currently, knitted fiber-reinforced composites are mainly used in applications involving substantial impact stress, such as helmets and car crash guards, but the applications of knitted fiber-reinforced composites can be expanded if its weak tensile strength is improved. Carabiners are one possible application with high potential and a lack of extensive previous research. Optimizing knitted fiber-reinforced composites with improved tensile modulus can help create lighter, but still structurally strong, carabiners. Climbing-grade composite carabiners are not currently available, but these lighter carabiners can help stem the many issues caused by the excessive weight on climbers' harnesses. To achieve this goal, we used computer-aided design and micro-scale simulations to optimize the design of the knitted-fiber reinforced composites to improve the tensile modulus to be comparable to standard carabiner materials.

## 1. Introduction

Knitted fiber-reinforced composites are characterized by extremely strong interlacing fibers knitted in various patterns. The two main types of knits are the weft and warp knits. Knitting horizontal rows one by one is considered the weft style of knitting, while knitting vertically and interlacing parallel rows is considered warp knitting. We focused on weft knits because their superior contourability is useful to account for the complexity of carabiner shapes [1]. As well as excellent contourability, knitted fabrics carry the advantages of high impact strength, low weight, and through-the-thickness strength that helps prevent delamination [2, 3]. These strengths make knitted fiber-reinforced composites a prime candidate for applications favoring low weight while maintaining strength, such as with carabiners.

Carabiners are widely used from rock climbing, sailing, construction, and even military applications. Carabiners are used to attach a climber's harness to safety ropes but can also be used to attach various tools and equipment on one's harness. Especially in the case of rock climbers, who often carry many various pieces of equipment on their harnesses, the cumulative weight of many metal carabiners can add up to a considerable weight. Carrying a heavy load is extremely dangerous for climbers who are often in dangerous and exposed positions high up. Having lots of heavy metal carabiners on one's harness can cause dangerous weight shifts, disrupt balance, and waste substantial amounts of energy. Decreasing the weight of carabiners can help lower overall carrying weight, which is a priority in climbing, to make rock climbing more energy efficient and safe [4, 5].

Some have attempted to decrease weight by creating polymer carabiners; however, polymer carabiners are too weak to use for climbing [6]. Composites are another promising alternative with light weight and substantially greater strength than polymers [7]. Additionally, the high impact strength of composites is beneficial because carabiners need to have high impact strength for if the carabiner is slammed against a mountainside or wall during climbing.

For its potential to decrease weight while offering many strength benefits, composites are a promising candidate for carabiner use; however, as far as the authors know, research done by Scott [4] is the only paper that has attempted to create a composite design for carabiners. His approach examined many different types of composites, materials, and knitting patterns for their viability in carabiners. This broad view is useful to holistically examine composites for their potential in carabiners, but there needs to be a more focused and detailed testing approach to optimize the use of specific composite types in carabiners [4]. Therefore, we are focused on optimizing, specifically, knitted fiber-reinforced composites for carabiner use in order to decrease weight while maintaining strength. Our optimization process is centered around improving the tensile modulus of a knitted fiber-reinforced composite model by adjusting structural parameters and incorporating additional structural modifications.

Through the use of various programming, computer-aided design (CAD), and finite element analysis (FEA) platforms, our contribution can be summarized as:

Optimizing the plain weft-knitted fabric loop structure to have improved tensile modulus and be comparable to commercial carabiner standards.

Setting the stage for incorporating composites into carabiners and decreasing carabiner weight drastically by 44%.

In the next section, the methodology of our research is explained. In Section III, results are presented and discussed, and lastly in Section IV, the findings are summarized with additional future work.

## 2. Predicting Mechanical Properties

The loop structure of weft-knitted fabrics is complex, which can lead to high computational costs. Thankfully, due to the patternability of knitted fabrics, we only need to model a representative volume element (RVE) to predict how an entire fabric structure would behave. Modeling a RVE allows us to lower computational costs while still accounting for the complexity of the entire fabric structure.

Before a RVE can be modeled and tested, an established geometric model must be used to output coordinate points for the RVE. From those coordinate points, a CAD platform is used to create a 3D visualization of the structure, which is then tested in an FEA platform to predict the mechanical properties of the structure. In the following subsection, the creation of the 3D structure of our composite is described.

### *2.1. Geometric Modeling of Plain Weft-Knitted Fabric*

Vassiliadis et al. [8] have created a geometric model for plain weft-knitted fabric that has improved on past models, so we used their methods to create our fabric structure.

The geometric model is based on a few main structural parameters: the course spacing $c$, wale spacing $w$, and the thickness of the yarn $D$. Variables other than $x$, $y$, $z$, $c$, $w$, and $D$ have their own separate pre-defined functions that ultimately contribute to the three main functions in the following column [8]. The model ultimately uses the following three main functions to output coordinate points $(x, y, z)$ for part ΣΜΚΛ of Figure 1. Part ΣΜΚΛ describes a quarter loop, which can be later duplicated and reoriented to create the structure shown by Figure 1. Figure 1 represents our eventual loop model, which consists of two interlocking half loops. These half loops serve as a RVE because an entire fabric structure is just composed of identical repeated interlocking loops.

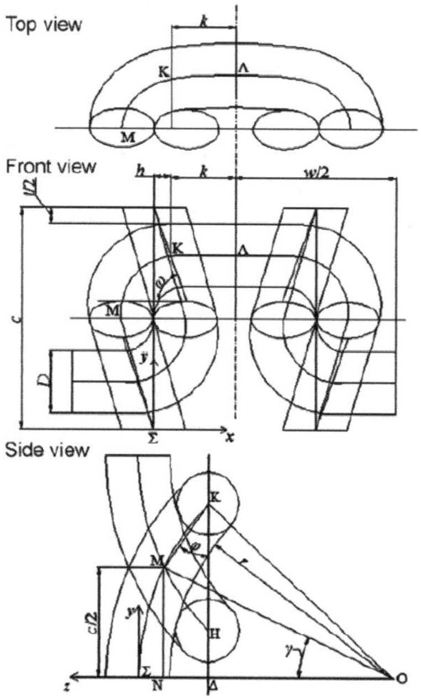


**Figure. 1**. Geometrical model of a plain weft-knitted unit cell. These diagrams illustrate the geometry of the interaction of two half loops. (c – course spacing; w – wale spacing; D – yarn thickness; t, r, Φ, γ, ω and h – parameters given in supplementary equations found in [8]; Σ, M, K and Λ – points on the curve that forms a quarter of a full loop) [8].

The following equations (1-6) describe the coordinates (*x, y, z*) of the three parts of ΣMKΛ:

Part ΣM $\left\{0 < y < \frac{c}{2}\right\}$:

$$x(y) = -\frac{D}{c}y \tag{1}$$

$$z(y) = \sqrt{(r + \tfrac{D}{2})^2 - y^2} - (r + \tfrac{D}{2}) \tag{2}$$

Part MK $\left\{\frac{c}{2} < y < \frac{c}{2} + R\right\}$:

$$x(y) = h - a\sqrt{1 - (\frac{y - \frac{c}{2}}{b})^2} \tag{3}$$

$$z(y) = \sqrt{(r + \tfrac{D}{2})^2 - y^2} - (r + \tfrac{D}{2}) \tag{4}$$

Part KΛ $\left\{x(y = \frac{c}{2} + R) < x < w/4\right\}$:

$$z(x) = OZ - \sqrt{A^2 - (x - OX)^2} \tag{5}$$

$$y(z) = \sqrt{(r + \tfrac{D}{2})^2 - (z + r + \tfrac{D}{2})^2} \tag{6}$$

These equations are inputted into a programming platform to output hundreds of coordinate points along the quarter loop ΣMKΛ. These coordinates are then imported into a CAD platform where the quarter loop is given a thickness, duplicated, and reoriented to create the interlocking half loop RVE.

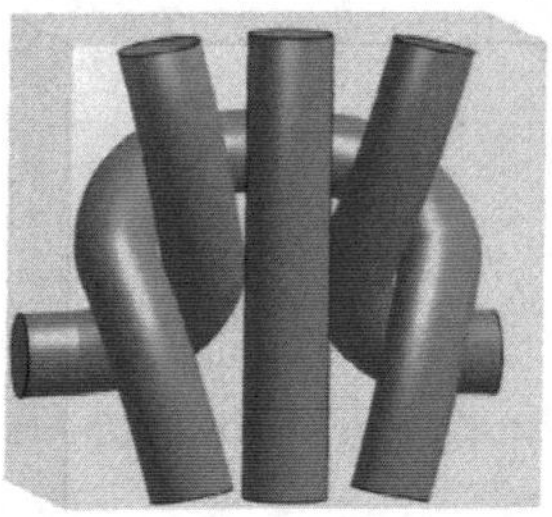

**Figure. 2**. Final CAD model of plain weft-knitted unit cell. This final structure includes a transparent encapsulating box, which is the matrix, and a vertical column in the middle, which is the reinforcing inlay.

Next, a matrix must be incorporated with the fabric to form a composite. The matrix must be tight-fitting in order to obtain accurate fiber volume percentages, so we surrounded the fabric by a tight-fitting box that represents the matrix. Finally, based on simulation results, additional designs, such as inlays were incorporated into the structure to reinforce certain aspects.

The final structure includes a vertical inlay to reinforce the strength of the composite in the wale direction. As for materials, we used carbon fibers and a Victrex PEEK 90HMF40 matrix. These materials were selected for their high tensile modulus, abrasion resistance, toughness, low density, and environmental durability [4]. The final model is as shown in Figure 2 above.

*2.2 Finite Element Analysis of the Unit Cell*

The 3D representative unit cell created in the CAD is then imported into an FEA platform to perform stress analysis tests. We tested the models in the wale (axial) and course (transverse) directions by fixing one side and applying a fixed displacement to the opposite side while calculating the volumetric stress as an output. Stress strain curves were then created based on the results of these tests.

We used initial results to modify the inlay design and thickness and to optimize fiber volume percentage. The highest-performing parameters were implemented in the final design as mentioned in the previous subsection.

Once the optimal composite design and parameters were determined, the optimal composite was then compared to standard carabiner materials subjected to the same loading circumstances. To account for variability in mechanical performance due to structural geometry, the aluminum and steel models were the same dimensions as the composite matrix. The full detailed results and analysis of the optimization of our composite model is presented in the following section.

## 3. Predicting Mechanical Properties

*3.1. Optimization of the Composite Structure*

Before our composite was compared to standard carabiner materials, it was first optimized. Additionally, comparisons were made to show the advantages of knitted fiber-reinforced composites over unidirectional (UD) fiber-reinforced composites and lone knitted fabrics.

First, an ideal fiber volume percentage was determined by tensile modulus, or stiffness. We tested 3.95%, the base model fiber volume; 7% and 11%, two intermediate values; and 20%, the maximum fiber volume percentage that the loop geometry allows space for. As shown in Figure 3, we determined the optimal fiber volume percentage with the highest stiffness to be the maximum fiber volume of 20%. This result was expected because the modulus of the carbon fibers is much higher than the modulus of the matrix, so increasing the fiber volume percentage increases the overall strength of the structure.

Next, the inlay structure was optimized. As shown in Figure 4, the knitted fiber-reinforced composite is stronger in the course, or horizontal, direction than the wale, or vertical, direction, so we added a vertical inlay that improved the strength in the wale direction by 2 GPa. We also found that the thicker the inlay, the better the stiffness of the fabric. To maximize the knitted loop yarn thickness and given the limited space with 20% fiber volume of the knitted fabric, we determined the optimal inlay thickness to be the same diameter as the knitted yarn thickness. Now, with this optimized model, it must be established that a knitted fiber-reinforced composite is superior to a knitted fabric without a matrix.

Making carabiners out of solely knitted fabrics is not feasible because the fibers would be susceptible to environmental factors and abrasions that could compromise the fibers' strength. A matrix must be incorporated with the knitted fabric to provide structural integrity, toughness, protection from harmful environmental conditions, and prevent friction between fibers.

Finally, as shown in Figure 4, in comparison to a UD fiber-reinforced composite, our knitted fiber-reinforced composite is weaker in the longitudinal, or fiber, direction but is stronger in the transverse direction. Despite being weaker in the longitudinal, or wale, direction, knitted fabrics have closer similarity in both the wale and course direction strength. The superior biaxial strength of knitted fabrics is beneficial for rock climbing where stresses are not always unidirectional. Additionally, knitted fabrics offer the benefits of increased impact strength and contourability. Based on these results, it can be established that our optimized knitted fiber-reinforced composite is superior to UD fiber-reinforced composites.

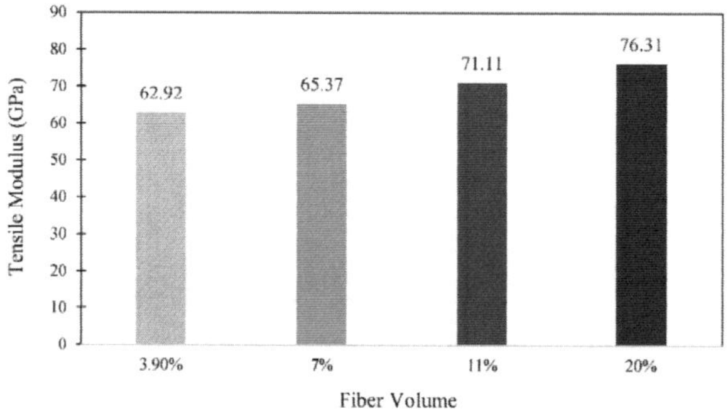


**Figure. 3**. Tensile modulus of various knitted fiber-reinforced composite fiber volume percentages. This shows that increased fiber volume leads to an increased tensile modulus.

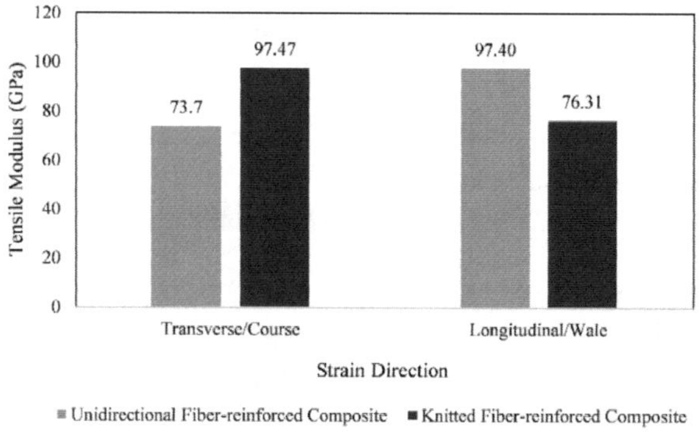


**Figure. 4**. Comparison of tensile modulus of UD and knitted fiber-reinforced composites in different directions shows the advantage of UD in the longitudinal direction and the advantage of the knitted in the transverse direction.

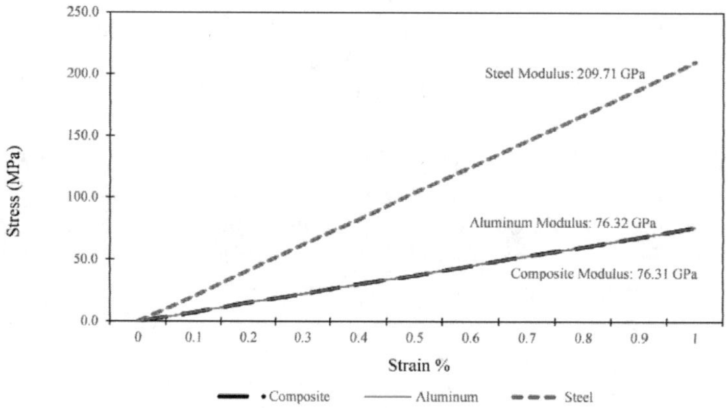


**Figure. 5**. Stiffness comparison of optimized composite and standard climbing-grade aluminum and steel. The labeled tensile moduli show the similar stiffnesses for our composite and aluminum.

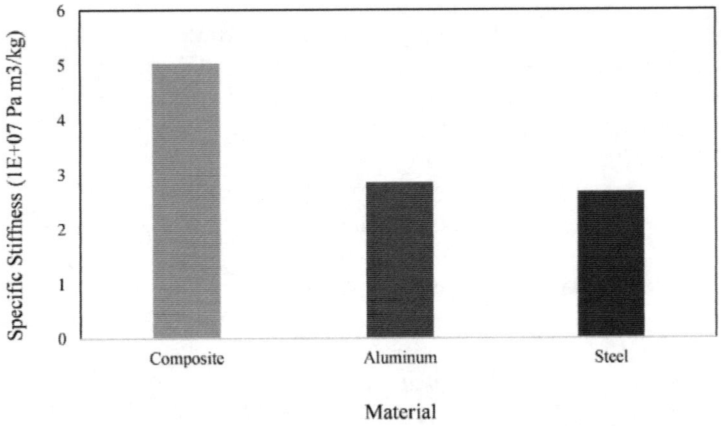


**Figure. 6**. Comparison of specific stiffness of different materials. Our optimized composite has the highest specific stiffness, which shows the superior strength to weight ratio of our composite.

### *3.2. Comparison to Standard Materials*
The final step is to gauge the viability of our optimized knitted fiber-reinforced composite by comparing it against aluminum 7075-T6 and steel, which are common climbing-grade carabiner materials. First, our composite was compared in terms of stiffness, or tensile modulus.

Figure 5 shows how our composite is very comparable to the stiffness of aluminum, which indicates that our composite should have adequate stiffness for carabiner use. Steel had a much higher stiffness than both aluminum and our composite. However, steel is almost three times heavier than both aluminum and our composite, so its strength to weight efficiency is inferior to the excellent strength to weight efficiency of our composite.

We quantified strength to weight efficiency by calculating specific stiffness, which is tensile modulus divided by the density of the material. Figure 6 shows that the specific stiffness of our composite is almost double the specific stiffness of both aluminum and steel, which demonstrates that our composite can greatly decrease the weight of a carabiner, approximately 44%, while maintaining strength.

## 4. Predicting Mechanical Properties
We have shown that knitted fiber-reinforced composites have the potential to drastically reduce the weight of carabiners and still maintain strength. We also found that the course direction of our composite design is the strongest, so for future reference, we recommend that the course direction of the composite is aligned with the longitudinal axis of a carabiner structure.

These results establish the promising potential of knitted fiber-reinforced composite carabiners, but many future steps must be taken before climbing-grade composite carabiners can be made commercially available. To start, we focused exclusively on plain weft-knitted fabrics, but many other knitting patterns should be investigated, such as milano, full-cardigan, and ribbed knits. Additionally, we only tested a representative structure in simulation, so the next step would be performing real physical tests, such as impact drop tests and stress tests that extend to fracture point, on a full carabiner structure. Finally, a viable and cost-efficient production process must be laid out to show the commercial practicality of composite carabiners.

There is much more work to be done, but with this paper providing an optimized knitted fiber-reinforced composite design, a solid foundation has been set with the hope that eventually knitted fiber-reinforced composites can make rock climbing more energy-efficient and safe.

IOP Conf. Series: Materials Science and Engineering 1044 (2021) 012008 doi:10.1088/1757-899X/1044/1/012008

## 5. References

[1] "Weft and Warp Knitting Differences", Machine Knitting Advice. [Online]. Available: https://machineknittingadvice.com/weft-and-warp-knitting-differences/. [Accessed: 10- Jul- 2020].
[2] N. Padaki, R. Alagirusamy and B. Sugun, "Knitted Preforms for Composite Applications", Journal of Industrial Textiles, vol. 35, no. 4, pp. 295-321, 2006. Available: 10.1177/1528083706060784 [Accessed 10 July 2020].
[3] S. Pandita, D. Falconet and I. Verpoest, "Impact properties of weft knitted fabric reinforced composites", Composites Science and Technology, vol. 62, no. 7-8, pp. 1113-1123, 2002. Available:
10.1016/s0266-3538(02)00057-x [Accessed 10 July 2020].
[4] V. Scott, "Design of a composite carabiner for rock climbing", Undergraduate, Imperial College London, 2008.
[5] M. Twight and J. Martin, Extreme Alpinism. The Mountaineers, 1999.
[6] "Carabiner, Polymer", ASP, Inc., 2020. [Online]. Available: https://www.asp-usa.com/products/carabiner-polymer. [Accessed: 10- Jul- 2020].
[7] D. Rajak, D. Pagar, P. Menezes and E. Linul, "Fiber-Reinforced Polymer Composites: Manufacturing, Properties, and Applications", Polymers, vol. 11, no. 10, p. 1667, 2019. Available: 10.3390/polym11101667 [Accessed 17 July 2020].
[8] S. Vassiliadis, "Geometrical modelling of plain weft knitted fabrics", Nopr.niscair.res.in, 2007. [Online]. Available: http://nopr.niscair.res.in/handle/123456789/406. [Accessed: 27- Jul- 2020].

IOP Conf. Series: Materials Science and Engineering 1044 (2021) 012009 doi:10.1088/1757-899X/1044/1/012009

# The BIM-based Integrated Design of the SHiP Project Decay Volume

**A Salzano[1,4,*], A Miano[1,4], G Porfidia[2,4], A Fiorillo[1,4], A Prota[1,4], R Jacobsson[3,4]**

[1] Department of Structures for Engineering and Architecture, University of Naples Federico II, 80125, Napoli, Italy
[2] Future environmental design (FED), SPIN OFF of University of Naples Federico II, 80125, Napoli, Italy
[3] European Organization for Nuclear Research (CERN), Geneva, Switzerland
[4] SHiP Collaboration

[*]Corresponding author: antonio.salzano@unina.it

**Abstract.** The Search for Hidden Particles (SHiP) experiment is a new general-purpose fixed target facility proposed at the CERN Super Proton Synchrotron accelerator to search for long-lived exotic particles associated with Hidden Sectors and Dark Matter. This paper reports on the BIM integrated design of SHiP's decay volume, a conical steel vessel under vacuum that should host several large particle physics detector systems. The use of BIM characterized the design of the decay volume, both in the modeling and structural design phase, and in the process definition phase for the realization and implementation in the facility of the device. This procedure helps to minimize the risks of incorrect design and construction of the device during the whole process. With the automation of the virtual model and the use of interoperable software, in addition to speeding up the exchange of information, it is possible also to export the detailed information of the structural design directly to the numerical control machines for the prefabrication of the various steel modules. Then, the BIM approach to support the integrated design of the SHiP project decay volume from the conceptual planning to the construction phase is shown in this work.

## 1. Introduction

The Search for Hidden Particles (SHiP) experiment [1-7] is a new general-purpose fixed target facility proposed at the CERN Super Proton Synchrotron (SPS) accelerator to search for long-lived exotic particles associated with Hidden Sectors and Dark Matter. This paper reports on the BIM integrated design of SHiP's decay volume, a >2000 m3 conical steel vessel under vacuum that should host several large particle physics detector systems. As part of the SHiP experiment, the use of BIM characterized the design of the decay volume, both in the modeling and structural design phase, and in the process definition phase for the realization and implementation in the facility of the device. The design of the decay volume has been characterized by the writing of a visual scripting algorithm in a visual programming environment (Dynamo, 8), which allows the creation of complex geometries. This algorithm allows passing from static to dynamic design, reducing the time for modeling and optimizing the model through a series of comparisons between the different design solutions. The data contained within Revit's [9] adaptive families, used to model the individual elements that make up the decay, have been exported with the use of a specific scripting algorithm for this operation. The structural analysis has been performed with Robot software [10], importing the geometric model of

Revit [9] into the software, transformed into a structural model using the Analyze nodes contained in Dynamo [8]. The realization of the decay volume, from the production in the workshop of the various panels to the assembly in the experimental area, has been divided into several work phases. During each phase, using the digital prototype, all the individual processes leading to the construction of the device were defined and designed, focusing on possible interferences that can arise during the whole process. To study this process, Navisworks software [11] has been used in order to combine the information from different software. The Navisworks software [11] also identified and corrected the interferences between the various models and those that can occur during the assembly operations of the structural elements of the device. The federated model in Navisworks [11] is used for sharing information and coordinating models with all stakeholders. This procedure helps to minimize the risks of incorrect design and construction of the device during the whole process. With the automation of the virtual model and the use of interoperable software, in addition to speeding up the exchange of information between the various work groups, it is possible also to export the detailed information of the structural design directly to the numerical control machines for the prefabrication of the various modules to the workshop. Then, the BIM approach to support the integrated design of the SHiP project decay volume from the conceptual planning to the construction phase is shown in this work.

## 2. Overview of the SHiP project

The SHiP research facility is composed of a large infrastructure with a dual detector system (Figure 1). The upstream system is specialised in the direct detection of Dark Matter and in performing measurements on neutrinos. The second detector system is dedicated to searching for decays of so-called Hidden Particles and allows probing a large variety of physics models with light long-lived exotic particles. It is based on a 50 m long decay volume followed by a spectrometer and by detectors for particle identification. A critical component of SHiP is the muon shield. It consists of a chain of magnets that deflect away from the detector the high flux of muons. The deflected flux of muons defines a conical region in which the decay volume and the detector systems may be placed. To suppress the background from neutrinos interacting in the fiducial volume, the decay volume must be maintained at a pressure of <10-3 bar by means of a vacuum vessel.

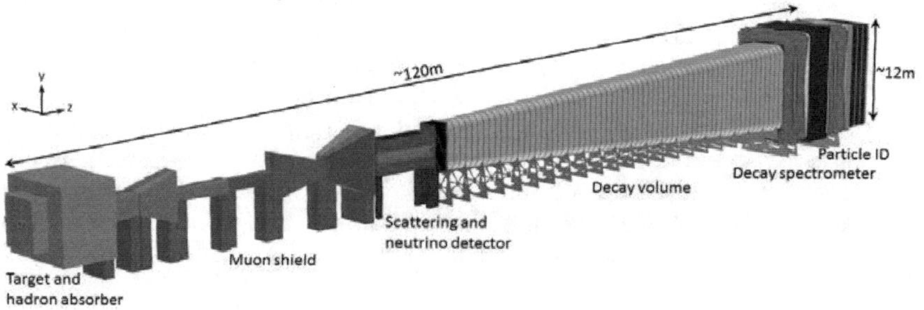


**Figure 1**: Overview of the SHiP experiment.

SHiP's vacuum vessel has a total volume of ~2040m$^3$ and logically consists of two parts, the volume in which a decay vertex is accepted, and the spectrometer section. The spectrometer section runs through the spectrometer magnet and lodges four tracker stations, which are symmetrically located with two stations upstream and two downstream of a large spectrometer magnet. An upstream and a downstream end-cap close off the ends of the vacuum vessel. To further ensure that signal candidates are not produced by neutrino or muon interactions in the upstream detector system or the decay volume walls, the decay volume is completely covered by a high-efficiency background tagger system, capable of detecting the charged particles produced in the interactions with the surrounding structure. The current baseline for the background tagger system is based on a "Liquid Scintillator detector" which fills the walls compartments of the decay volume. The dimensions of the experimental area (Figure 2) have been determined starting from the dimensions of the whole SHiP Detector.

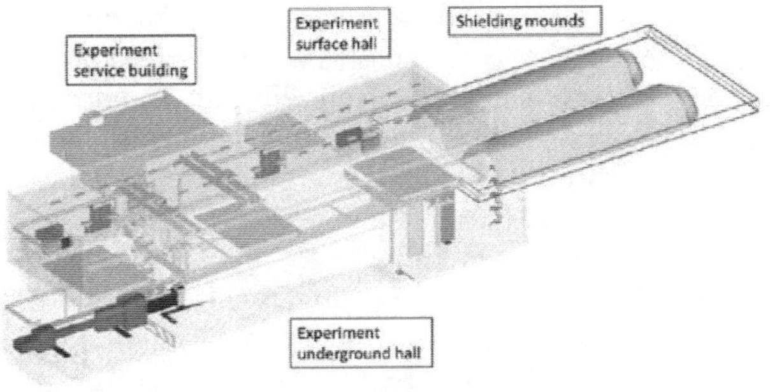


**Figure 2**: The experimental area.

The total length of the detector is around 120 meters. The assembly of the entire device determines the arrangement of the ground floor, access doors and cranes that will be used for handling loads. The dimensions of the opening on the floor have been chosen in agreement with the maximum dimensions of the individual elements of the device to be lowered. The decay volume will be pre-assembled on site in few big pieces, that will be then lowered in the underground experimental area, where have to be welded among them and connected with the other subcomponents.

## 3. Methodology

The whole methodology adopted is presented herein. With the use of Dynamo [8] and Revit [9] software, the decay volume has been modeled and imported into Robot [10] to carry out the structural analysis and the safety checks. Once the model has been verified, it has been re-imported into Revit [9] for the export of geometric data through the use of a Dynamo [8] script.

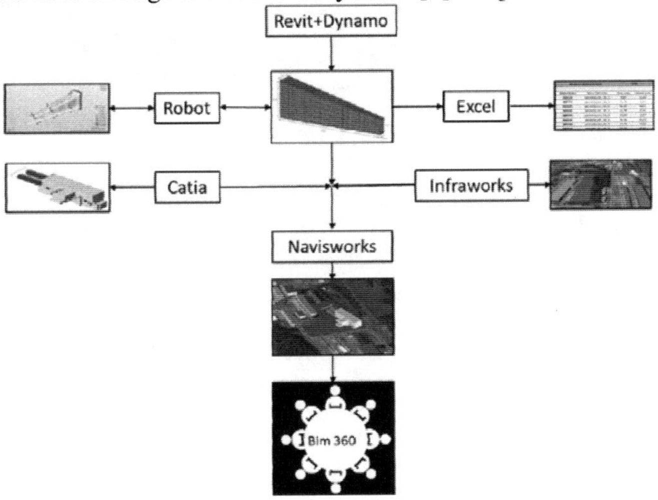


**Figure 3**: Flowchart of the BIM process.

The decay volume model made with Revit [9], the experimental area model made with Catia [12] and the terrain and infrastructure surrounding the experimental area model made in Infraworks [13] were then federated into Navisworks [11]. The complete model has been used to study the assembly phases of the elements that make up the decay volume. The federated model is shared on a BIM 360-type platform, with which the various stakeholders can intervene throughout the device design and construction process. The described procedure is summarized in the flowchart presented in Figure 3.

## 4. BIM integrated design of the decay volume

### *4.1. Integrated design development of the decay volume*

Considering the goal of minimising the masses and the constraints on the geometry, the steel structure has been chosen. The main conditions to consider in the design process were: a) the necessity to allocate space for the liquid scintillator inside specific compartments inside decay volume; b) the necessity of reducing the thickness of the steel plates as much as possible. Based on these conditions, a box section of S355JO (J2/K2)W Corten steel has been selected. The decay volume has been designed inside a BIM framework, using Revit [9] modeling software and using the open source visual platform, Dynamo [8], integrated directly into Revit [8]. The reason for this choice is that the BIM computational design allows designing highly flexible parametric geometries. The parameters are always editable within the chain of nodes in which they are structured. Any changes to these parameters are directly reflected on the geometry. The Dynamo [8] work environment is structured with nodes and relationships; the nodes manage specific functions and are connected to the other nodes, composing an uninterrupted flow of data (e.g. see Figure 4a). Writing this algorithm avoids the manual insertion of all the elements that form the decay volume, reducing the time for modeling. With this modeling approach, the accuracy of the model increases and the modeling errors are reduced. Finally, by changing the input parameters, it is possible to design structures with different geometries to optimize the structural model. The box section is made up by the following members:

- Inner and outer steel sheets with a thickness of 20 mm (see Figure 4b);
- Transversal stiffening members spaced by 800 mm, with a thickness of 10 mm and a height of about 400 mm (see Figure 4c);
- Longitudinal stiffening elements, with a thickness of 10 mm and with an interspacing varying from 804 mm to 1702 mm (see Figure 4d).

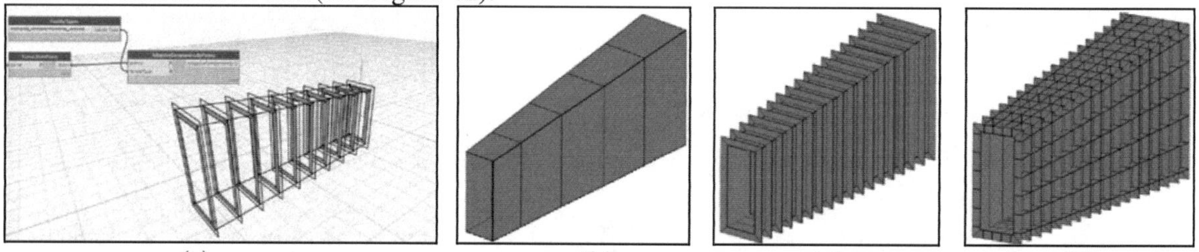

(a) (b) (c) (d)

**Figure 4**: a) *Adaptive.Component.ByPoint* node for the insertion of adaptive families within the generated surfaces b) Inner steel sheet; c) Transversal and d) Longitudinal stiffening members.

The model generated in Revit [9] through the use of Dynamo [8] software has been imported into Robot [9]. Then, the structural calculations have been performed with Robot [9]. Finally, the data contained within the elements of the model generated in Revit [10] were exported using a Dynamo [8] script. The script allows to export the output data and report them directly on an excel [14] type spreadsheet format. The various elements contained within the Generic Models category have been called up using the All Element of Category node. They have been connected with a series of nodes to know the surface, the volume, the element ID and the name. All this information has been sent to the *Data.ExportExcel* node through the List Creates node. Then, the *Data.ExportExcel* is used to write information in an excel [14] sheet.

### *4.2. Integrated constructive design of the decay volume*

The presented study focuses also on a careful assessment of the assembly phases in the factory and situ. The fundamental principle is to give preference to prefabrication in the factory, a place equipped with multiple equipment, production machines and processing control protocols. This choice is assumed to allow the reduction of the imperfections/errors in the assembly phases on site. Then, the proposed fabrication processes are divided into two phases. The first one concerns the prefabrication in the factory, while the second one is all the on-site assembly processes through manual processing.

Regarding the factory production phase, a basic panel with a width of 2.4 meters with variable height will be produced in the factory workshop, using numerical control machines with thermal break. The welds will be carried out following specific procedures and sequences. During the assembly phase of the individual panels in the workshop, any deformations will be corrected by means of special procedures ("localized heating" and similar procedures) in order to guarantee the most severe execution class (EXC4) foreseen by UNI EN 1090-2 [15]. As a starting point, the various elements that make up the decay volume consist of an internal panel, a series of vertical and horizontal stiffeners and external panels for closing the individual cells for containing the liquid scintillator. The realization of the single components will be carried out through the use of numerical control machines, using the output data that can be obtained from the parametric objects that make up the 3D model of the decay volume, extrapolating the necessary information with the use of a special algorithm created in Dynamo [8]. The assembly phase involves first joining the vertical stiffeners, previously subjected to chamfering, on the internal panel by means of welds with complete penetration with filler material. The next step consists of inserting and welding the horizontal elements, which define the individual cells. The welds will be completely penetrated with filler material on elements subjected to chamfering. Finally, the external panels (Figure 5) are welded.

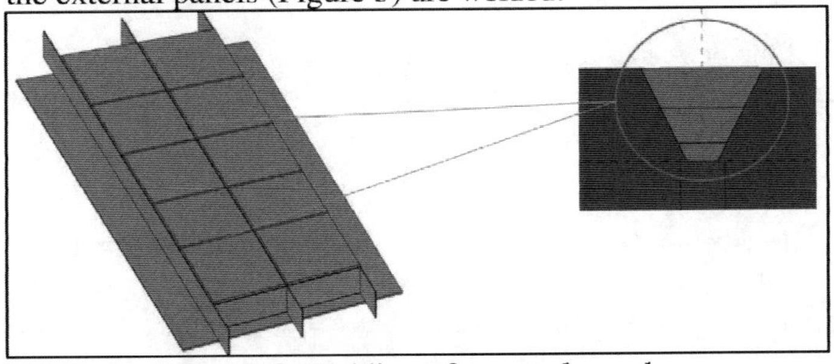

**Figure 5**: Welding of external panels.

Regarding, the in-situ assembly procedures, once made, the 2.4-meter basic panels will be transported to the construction site to be assembled into a single panel consisting of a maximum of 6 modules. The decay has been divided into 4 blocks to minimize the number of welds to be made within the underground area, where the experiment has to be done. This subdivision into macro blocks allows increasing the flatness between the various panels. Figure 6 shows the main phases that lead to the complete union between the two panels.

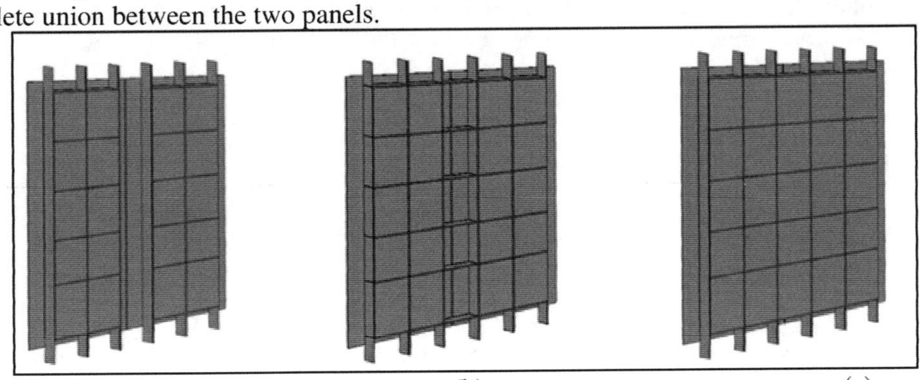

(a) (b) (c)

**Figure 6**: a) Internal welding; b) Horizontal stiffening; c) External panels.

Then, it is crucial to plan the mounting sequence underground. After the transport within the experimental area by mechanical means and the movement with a bridge crane from the first floor to the basement, finally the panel will be positioned at the planned point, using slides. The whole process has been simulated with Navisworks [11] to identify the possible interferences that can occur during the entire construction phase of the device. The panel is transported within the experimental area by means of a mammoth type mechanical device. Arrived in the predetermined position, it is first raised, then moved to the opening that connects the ground floor with the basement and finally is lowered and positioned on special slides to be brought to the final position where it will be mounted. The most important operations are the transport within the experimental area, the handling of loads with the use of the bridge crane and the insulation of the panel on the detective SHIP (Figure 7).

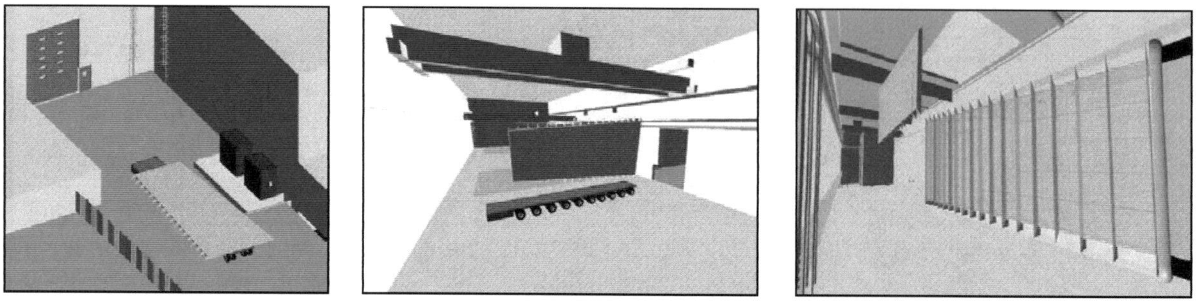

(a) (b) (c)

**Figure 7**: Assembly of the decay volume in Autodesk Navisworks software [9]: a) panel transport in the experimental area; b) panel lifting by means of a bridge crane; c) panel assembly.

## 5. Conclusions

The SHiP experiment is a new general-purpose fixed target facility proposed at the CERN to search for particles associated with Hidden Sectors and Dark Matter. This paper reports on the BIM integrated design of SHiP's decay volume. The use of BIM characterized the design of the decay volume, both in the modeling and structural design phase, and in the process definition phase for the realization and implementation in the facility of the device. This procedure helps to minimize the risks of incorrect design and construction of the device. With the automation of the virtual model, it is possible to exchange information and to export the design to the numerical control machines for the prefabrication of the various steel modules. In conclusion, the BIM approach supports the integrated design of the SHiP project decay volume from the conceptual planning to the construction phase.

## Acknowledgments

The authors would like to thank CASTALDO SpA for the valuable inputs onthe welding techniques and the assembly procedures. Moreover, the authors would like to thank Dr. Pablo Santos Diez for the valuable inputs on experimental area description and models.

## 6. References

[1] SHiP collaboration 2013 Proposal to search for heavy neutral leptons at the SPS. arXiv preprint arXiv:1310.1762.

[2] SHiP collaboration 2019. Sensitivity of the SHiP experiment to heavy neutral leptons. *Journal of High Energy Physics*, **4**, 77.

[3] SHiP collaboration 2019. Fast simulation of muons produced at the SHiP experiment using Generative Adversarial Networks. *Journal of Instrumentation*, **14**(11), P11028.

[4] SHiP collaboration 2019. The experimental facility for the Search for Hidden Particles at the CERN SPS. Journal of Instrumentation, **14**(03), P03025.

[5] SHiP collaboration 2020. Measurement of the muon flux from 400 GeV/c protons interacting in a thick molybdenum/tungsten target. *The European Physical Journal C*, **80**, 1-11.

[6] SHiP collaboration 2020. The Magnet of the Scattering and Neutrino Detector for the SHiP experiment at CERN. *Journal of Instrumentation*, **15**(01), P01027.

[7] Miano A, Fiorillo A, Salzano A, Prota A, Jacobsson,R 2020. The structural design of the decay volume for the Search for Hidden Particles (SHIP) project. *Arch Civil Mech Eng-Under Review*.

[8] Dynamo (2021), Autodesk Dynamo, Autodesk.

[9] Revit (2021), Autodesk Revit Architecture, Autodesk.

[10] Robot (2021), Autodesk Robot, Autodesk.

[11] Navisworks (2021), Autodesk Navisworks, Autodesk.

[12] Catia (2019), Dasault Systemes, developer Autodesk.

[13] Infraworks (2019), Autodesk Infraworks, Autodesk.

[14] Microsoft Project 365, developer Microsoft Corporation.

[15] EN 1990 (2002): Eurocode 0: Basis of structural design, Brussels, Belgium.

IOP Conf. Series: Materials Science and Engineering 1044 (2021) 012010 doi:10.1088/1757-899X/1044/1/012010

# A Digital Logbook as an Interactive Tool to Fulfil Service Companies' Needs and Requirements in Building Renovations

**Martina Signorini[1,*], Sonia Lupica Spagnolo[1] and Mario Claudio Dejaco[1]**

[1]Politecnico di Milano, Department of Architecture, Built environment and Construction engineering, Via G. Ponzio 31, 20133 Milano, Italia

martina.signorini@polimi.it

**Abstract.** The lack of an appropriate information management along the life cycle of a construction work is one of the major issues that AEC sector is facing. The collection and exchange of information is already complex for new buildings, but it becomes even more difficult for renovations. Thus, the building logbook could have an essential role in improving the availability and quality of data for several stakeholders involved in the building process. The repository of building-related information, such as energy consumption and production or maintenance operations, is a bilateral tool to connect users and third parties. The aim of the paper is to present the digital logbook developed within BIM4EEB, an ongoing Horizon2020 project that has analysed firstly the needs and requirements necessary for its development and then it has defined its structure in order to be stored and accessed inside the BIM management system.

## 1. Introduction

Buildings are among the biggest consumers of energy in the world. In fact, it is estimated that in many International Energy Agency (IEA) member countries not only the AEC sector is responsible for over 40% of primary energy consumption, but also that this energy demand doubled between 1971 and 2010, guided primarily by population increase and economic growth. Considering buildings number progressive growth, their global energy demand is expected to increase by a further 30% by 2035. In countries where current buildings will remain in place for years, the renovation should be considered the best way to focus on, through implementation of energy codes and minimum performance standards in existing buildings [1]. To meet both the EU 2020 targets and the commitment undertaken in Paris in December 2015, it is crucial to increase the current EU renovation rate from 1.2 % per annum to 2–3 %. Despite building renovation feasibility from an economic and technical point of views but also social and environmental benefits that could come from, the amount of renovations is still low and under the expected level. Building owners and stakeholders encounter numerous barriers to improve the energy performance of their buildings [2].

Several kinds of barriers have to be taken into account in renovation interventions, as financial issues, working time control, discomfort of users and a difficult decision making among different suitable interventions. Difficulties in accessing information concerning the facility, especially when such information is mainly in a paper-based format, are a common barrier for stakeholders. Moreover, during operation and maintenance, the involved stakeholders have sometimes to deal with as-built data that are out of date or with a lack of knowledge about building components and systems that make the choice of the best alternatives for renovation difficult. In some types of renovations, a large number of stakeholders are involved, and this complicates the exchange of information and it results in bad communication. In addition, financial and legal difficulties have to be taken into consideration.

Energy Performance Certificates (EPCs) could be useful to solve these obstacles, but they have limitations. In fact, they don't give tailor made information about renovation and they don't communicate increased thermal comfort and air quality, higher levels of natural lighting and improved health of occupants that are some of the most important benefits and drivers for renovation.

Consequently, Buildings Performance Institute Europe (BPIE) proposes to expand EPCs into Building Renovation Passports (BRPs), a user-friendly long-term step-by-step renovation roadmap adopted by owners to plan renovations and to collect all relevant information about a specific building and that could be used in combination with the so called logbook [3]. The concept of logbook or building passport born as collector and repository of building information, containing significant data about renovation intervention. Thus, it became not only a fundamental source of information for stakeholders, but it is a beneficial instrument for owners that could improve the knowledge about their buildings and consequently the quality of them [4][5].

Within this context, the BIM4EEB project has developed a structure for digital logbooks [6][7], as an archive of building related information in a digital format coupled with a renovation roadmap within iBRoad project [8][9], with the aim to guide building owners through their building renovation process by giving a customised renovation plan.

The use of digital logbooks has several benefits, for instance, the possibility for building owners to access important data, to do an in-depth building diagnosis, to receive alerts and reminders, to make useful comparisons with other buildings, to read the detailed version of their renovation roadmap, and to assess relevant financing options available for the specific renovation.

Besides the building owners, inhabitants and authorized third persons – such as energy experts, utility companies, facility managers, financial institutions and public authorities – are deemed stakeholders who can contribute or benefit when accessing the logbook.

The lead consultant/designer, responsible for developing the building logbook, provides benefit ensuring that the client's requirements are satisfied throughout the process and that the actual design intent is passed to the facility manager. Facility managers can have firstly a better understanding of the building and they can contribute in tracking changes, furthermore they may have ready access to information on the design, commissioning and energy consumption allowing a better management and possible improvements in energy efficiency. Occupants or inhabitants may understand better how to use spaces and to improve comfort and energy efficiency. Eventually, detailed information about the real time status of the building, materials and systems and energy consumption is needed by financial entities and real estate professionals.

Once defined on one hand the barriers that stakeholders have to overcome during renovations in order to obtain an improvement of building conditions and on the other hand the several benefits that each stakeholder gains from the digital logbook, the first step in order to develop a digital logbook is the definition of needs and requirements.

## 2. Research and methodology

BIM4EEB, a current Horizon2020 project founded by European Union, promotes BIM-based renovations through an impressive toolset that is capable to support designers, construction companies and services companies respectively in design and planning phase, to perform efficiently the work and to provide interesting solutions for building renewal. Therefore, the project is creating a BIM management system with specific tools for quickly mapping buildings, for improving HVAC design, operation and management, and for better tracking renovation operations. To define which information should be collected within the BIM4EEB digital logbook, the paper analyses existing logbooks because they have different developments and also different name, purpose, concept, and features across European countries. The various European initiatives on such tool will be firstly examined. The second part of the chapter will be instead reserved to the investigation of the methodology adopted to create the digital logbook.

*2.1 State of the art*

The table below presents a series of European measures on building passports and logbooks.

**Table 1.** Building passport / logbook overview

| Name | Country | Structure/Documents |
|---|---|---|
| **Building folder (Hausakte)** | Germany | The building folder is characterized by two parts: 1) the Building Certificate and 2) the "House Document". In 2004, the digital version of the building folder was developed in 6 modules: 1. Characteristics; 2. Description; 3. Address; 4. Summary tables; 5. Drawings, calculations, photos; 6. Documents and contracts [10]. |
| **As-Built File (Oplever-dossier)** | Netherlands (NL) | Opleverdossier (As-Built file) is a record of information on a residential building giving details about the technical quality of that building and guidance on maintenance. It has been introduced in the Building Quality Law in April 2016. Collected information relate to the completion of the building, building maintenance and servicing, moisture and ventilation, construction and maintenance of heat insulation, mechanical ventilation, electrical installation drawings, pipework and sewerage drawings, information on materials used, and other plans and drawings [11]. |
| **Fascicolo del fabbricato** | Italy (IT) | It is a technical document containing all the distinctive elements of a building - identification, design, structural and plant information - with the aim of having a complete knowledge and consequences of prevention and safety. The document is voluntary and was developed in 1999. Thanks to the D.P.R. 380/2001 and the new technical standards for buildings - D.M. January 14, 2008 – the Italian logbook spread more widely, and its use increased. Municipalities, provinces and regions have declined it in different ways, but the key contents are the same. An example of contents is the following: records in relation to the location of the property in the territory, cadastral card and size of the property; techno-typological data (to be updated with information on the evolution of conditions of the building and its parts); data on materials and components used; data on changes and adjustments (i.e. structural and plants systems, intended use) [12]. |
| **Building Logbook** | UK | Generally, in UK reference is made to CIBSE Building logbook. It offers a standard format covering how a building is intended to work and how it is meant to be maintained and serviced. It provides a mean to record the energy use and maintenance of the services within the building and it also helps with monitoring and maintaining occupant satisfaction by keeping a log of indoor environmental quality (IEQ) related complaints and the response actions. This tool collects information grouped into: Building history; Purpose and responsibilities; Links to other key documents; Main contacts; Commissioning, handover and compliance; Overall building design; Summary of areas and occupancy; Summary of main building services plant; Overview of controls/BMS; Occupant information; Metering, monitoring and targeting strategy; Building performance records; Maintenance review; Results of in-use investigations [13]. When speaking about UK building logbook, it is important to mention Home information pack (HIP), containing all the documents usually required when buying and selling a home, valid in UK and Wales [14]. In addition, also the building logbook used in Cornwall, a summary document describing the new or restored building and how it works, can be considered [15]. |

*2.2 Case study: the digital logbook within BIM4EEB project*

The first step to build the logbook is the definition of stakeholders needs and requirements. Since the EU project involves partners from different countries (Italy, Finland, Spain, Sweden, Ireland, Cyprus, Germany, Belgium, and Poland), national differences were taken into account because of the influence on needs and requirements. Beside different national renovation standards and laws, conditions of building stock could vary together with calculation methods for assessing energy performances, market situation for energy audit products and funding programmes and national incentives.

The following questions were considered to investigate and collect useful data for service companies:

- Which are the services to be taken into consideration?
- Who will be the main users of digital logbook?

- Which are the main interests of service companies in renovation processes?
- Which are the main sources of information?
- Which kind of information need to be stored in a digital logbook?

Specifically, energy analysis and audits, energy management, project design and implementation, maintenance and operation, monitoring and evaluation of savings, property/facility management, energy and/or equipment supply, provision of service (space heating, lighting, etc.) are the considered services within the project and about which information are collected.

The main stakeholders and users as well as their benefits derived from the digital logbook have been explained in the first chapter and are reported in a comprehensive outline in the figure below.

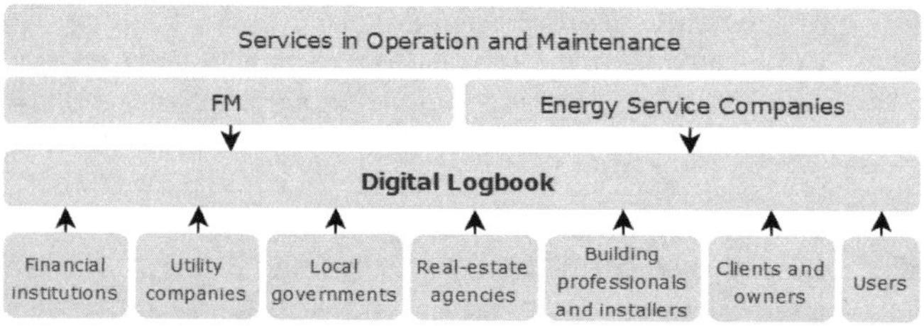


**Figure 1.** Stakeholders using digital logbook and taking advantage of it.

About the major interests of service companies in renovation processes several aspects must be considered as renovation policy impact, impact on energy demand, investment requirement, impact on comfort, impact on real-estate value, environmental protection policy, $CO_2$ savings, recommended timeline, health and safety, fire safety issues, connection with building automation system, visualisation such as work orders, how and where maintenance actions will be done, installation instruction, etc.

According to the literature, the building logbook should contain data such as building type and age, property plans and obligations, periodic synthesis data resulting from sensors continuously monitoring indoor climate (air temperature, air relative humidity), indoor air quality ($CO_2$ concentration, VOC content) and thermal comfort (PPD and PMV), information about energy consumption and energy cost, main significant interventions of executed maintenance (e.g. boilers replacement), etc. However, compared to the literature the BIM4EEB logbook concept has a different view of users and their access to it, factor that makes it unique in comparison to the other developed logbooks.

In fact, within BIM4EEB the logbook would become a tool for in-depth information on building relevant data and actions taken on it available to facility managers, but also to owners and clients, public authorities, occupants, designers and installers through the BIM Management System (BIMMS).
In this way building owners could be supported in investigating renovation options. Other stakeholders could access the same information for improving building performances or for updating requirements during building lifecycle.

The digital logbook is not only a management tool but it is also characterised by an operating feature: building operators could access the digital logbook and contribute actively to it reporting - if maintenance interventions for instance are considered - the most important actions, reserving the other executed maintenance operations for the planned O&M program.
Several kinds of data, which have to be updated during building lifecycle, are collected regarding different sources of information as building permits, properties, localization and legislation; agreements between stakeholders; agreements related to building use and operation; reports and certificates or data directly collected around the building.

## 3. The structure of digital logbook

The digital logbook has been structured with the aim to overcome issues related to the lack of information and partial knowledge about building conditions. The developed structure follows a certain flexibility to serve both national/regional needs but also to be integrated into a wider European approach.

To obtain the final structure of the digital logbook, the guidelines of the renovation roadmap managed by iBRoad project have been partly followed, but, unlike the iBRoad renovation roadmap, the digital logbook has been conceived as a synthesis of the most relevant building information from several points of view and not as a customised and long-term renovation plan. Even if, in this moment, it is a mainly a descriptive repository, in the future it can be extended with an evaluation of the current state of the building in the form of score given by the stakeholders themselves in order to have a major and numerical feedback.

The proposed BIM4EEB logbook framework is subdivided into five macro areas to facilitate the distinction between different topics. The first group contains the *general and administrative information* about the building and its units, users, and owners. In the *Building construction information* group, there are all the technical information and data about the building asset and its components. Information related to energy consumption and performance belong to the group *Building energy performance*. Maintenance activity for the building and its parts are collected within the *Building operation and use* group. Lastly, data connected to the building or its surrounding infrastructure through information and communication technologies (ICT) are placed in *IoT information*.

In addition, it is fundamental to define different levels to classify the information on the type of collected data. Therefore, the digital logbook is characterised by six fixed levels of information, as follows: *Subgroup of information*; *Information*; *Source of information*; *Stakeholders providing the information*; *Stakeholders requiring the information*; *Use of information in O&M*.
Since the information content for each category is very considerable, it is not possible to show in this scenario the whole developed logbook. Consequently, one of the most significant section related to the facility management has been represented in the figure below.

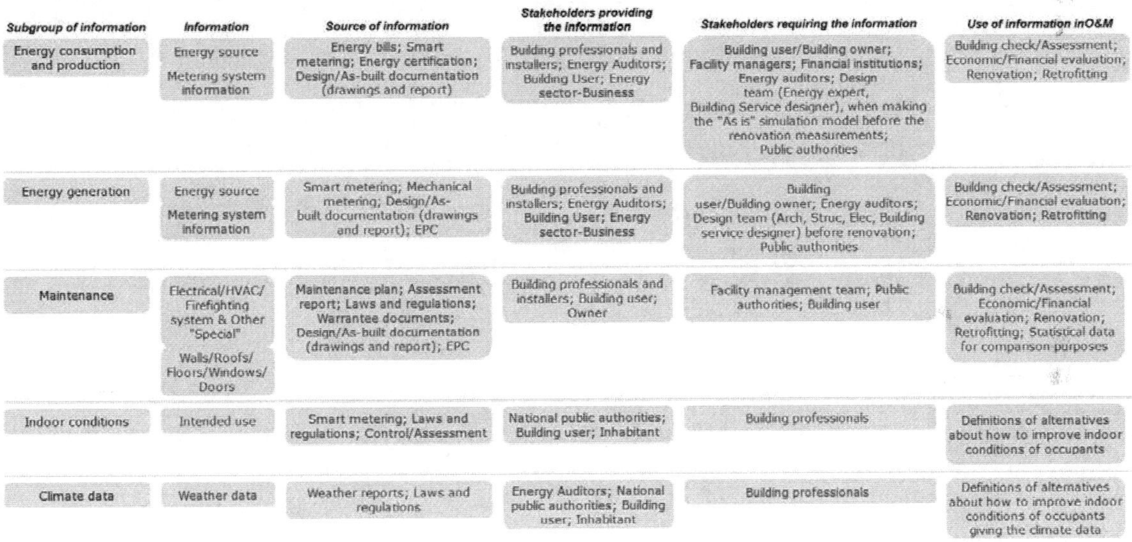


**Figure 2.** BIM4EEB digital logbook framework concerning building operation and use

## 4. Conclusions

The paper has been developed in the context of the EU founded project BIM4EEB. It proposes the structure of a digital logbook to overcome communication and information management issues that are typical in renovations.

Starting from an overview on the building logbook concept in Europe and focusing then on the digital logbook developed by the BIM4EEB project, some important conclusions can be outlined:

- monitoring regularly data collected from buildings, it is possible to reduce the divergence between building operation and building design
- inhabitants and occupants could understand better which is the design intent and buildings performance, their comfort, satisfaction and productivity could improve too
- data from digital logbooks can also improve the commissioning and handover process
- a better management and operation of buildings results in lower costs and in a reduction of $CO_2$ emissions.

**Acknowledgments**

The BIM4EEB project has received funding from European Union's H2020 research and innovation programme under grant agreement N. 820660. The content of this publication reflects the authors view only and the Commission is not responsible for any use that may be made of the information it contains.

## 5. References

[1] International Energy Agency 2013 Modernising building energy codes to secure our global energy future

[2] Fabbri M 2016 Understanding building renovation passports: customised solutions to boost deep renovation and increase comfort in a decarbonised Europe *ECEEE Summer Study on energy efficiency: Consumption, efficiency and limits* pp 1409–1417

[3] Buildings Performance Institute Europe (BPIE) 2016 Building Renovation Passports – Customised roadmaps towards deep renovation and better homes

[4] Lützkendorf T and Lorenz D 2011 Capturing sustainability-related information for property valuation *Build. Res. Inf.* **39** 3 pp 256–273

[5] Klomp B 2006 Improvement of information for owner-occupiers about the quality of their house *Home ownership in Europe: policy and research issues* pp 1-11

[6] BIM4EEB 2019 BIM4EEB: BIM based toolkit for Efficient rEnovation in Buildings [Online] Available: https://www.bim4eeb-project.eu/

[7] BIM4EEB 2019 D2.4 List of service companies' needs and requirements for BIM-based renovation processes

[8] Buildings Performance Institute Europe (BPIE) 2018 The Concept of the Individual Building Renovation Roadmap An in-depth case study of four frontrunner projects

[9] Buildings Performance Institute Europe (BPIE) 2018 Understanding potential user needs - A survey analysis of the markets for Individual Building Renovation Roadmaps in Bulgaria, Poland and Portugal

[10] European Construction Sector 2016 Policy measure fact sheet Germany Hausakte ('Building folder') Thematic Objective 1 pp 2001–2003

[11] European Construction Sector 2016 Policy measure fact sheet Netherlands Opleverdossier (As-Built File) Thematic Objectives 1 & 3

[12] Gruppo di lavoro del Consiglio Nazionale dei Periti Industriali e dei Periti Industriali Laureati (CNPI) 2017 Il fascicolo del fabbricato

[13] CISBE 2006 Building log book

[14] Office of the Deputy Prime Minister 2005 The home buyer and sellers guide to the home information pack

[15] Cornwall Council 2012 Building Standard –The Building Log Book pp 1–11

# Upgrading and Renovating Practice of the Schoolhouse in the High-density Old Town

**Zhiyuan Chen[1], Zhiyao Wang[1], Haiyang Qin[2,*]**

[1] Huangpu Education Bureau, Shanghai, 200025, China
[2] Tongji University, Shanghai 200092, China

*973455203@qq.com

**Abstract:** In view of the degradation of the existing structure of the school buildings in the old city and the need to update the education model in the new period, whether to upgrade the function of the school building has become a key issue for the education department to improve the quality of teaching. In the process of renovation, there are many challenges in the design and construction of how to upgrade and renovate the school buildings with excellent historical buildings with a long history in the old city with dense buildings and public facilities. This paper takes the renovation of excellent historical buildings at No. 95 Huqiu Road, Huangpu District, Shanghai as an example. Through the description of the buildings and implementation process, it highlights the important role played by technological innovation in the upgrading and renovation of school buildings.

## 1 Introduction

Huangpu District in Shanghai is the oldest urban area with the highest population density and the oldest history. It belongs to the high-density old urban area. The basic education facilities in this area are the most dense and outdated (excluding those that have been updated), 80% of which are included in historical buildings or excellent historical buildings or key cultural relics protection buildings. Figure 1 shows the distribution of basic education facilities in the area.

**Figure 1.** Distribution map of basic education facilities

The current "Internet+" model has accelerated social development, and the cultivation of innovative and creative talents is a country's core competitiveness.

In 2017, in order to deeply implement the national innovation-driven development strategy, efficiently promote the comprehensive reform of Shanghai education, and greatly complete the exploratory reform of emphasizing middle school teaching features, cultivating international innovative talents, and serving Shanghai as the key position for building a globally influential science and innovation center, the Huangpu District Education Bureau is in the process of transforming basic education facilities, innovation and pragmatism are guided by the school tenet of "different from the tradition, with a focus on the future". At the same time, the school buildings are functionally upgraded and transformed in compliance with the protection regulations of excellent historical buildings.

In the process of renovating school buildings, there are many problems need to be solved. For example, how to repair and improve the functions of the school buildings listed in the protection area in the high-density old city, how to deal with the relationship between the new and the old, how to highlight the cultural and special status of the original building, and how to create a innovative space that meet the requirements of safety and modern creativity training.

This article combines a new teaching model and takes the renovation of the old school building at No. 95 Huqiu Road, Huangpu District as an example. It focuses on the three aspects of architecture, structure and construction to explain how to upgrade the function of the school building and protect the cultural relics in the dense old city.

## 2 Function upgrade and renovate

### *2.1. Background*

The school building at No. 95 Huqiu Road is the original 3-storey school building, located west of Huqiu Road and south of Hongkong Road in Huangpu District, Shanghai (as Figure 2 shows). The north side of the school building is Hongkong Road, the west side is adjacent to the current 6-storey building of Shanghai Industrial and Commercial College, and the south side is adjacent to the current Huqiu residential building, the east side is Huqiu Road, and the east side of Huqiu Road is the 2nd phase of the 174 block project under construction.

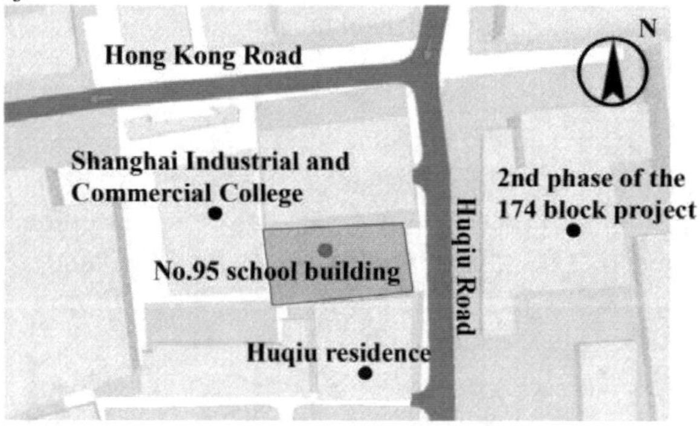


**Figure 2.** Schematic diagram of plane position

The school building was built in the 1920s, and the main structure of the house is a four-layer mixed structure. The seismic fortification intensity of the building is 7 degrees, and the earthquakes design are of the second group; the site belongs to the general earthquake-resistant level. The total length of the building is 35.18 m, the total width is 15.2 m, the total height is 22.5 m, and the total construction area is about 2502m$^2$.

The first and second floors mainly use the mixed load-bearing structure of brick wall, wood column and wood floor. The third floor mainly use the mixed load-bearing structure of brick wall, steel column, and concrete cast-in-place floor. The fourth mainly use brick wall bearing. The roof is mainly supporting purlin system of wooden triangular roof truss. The structure diagram is shown in figure 3.

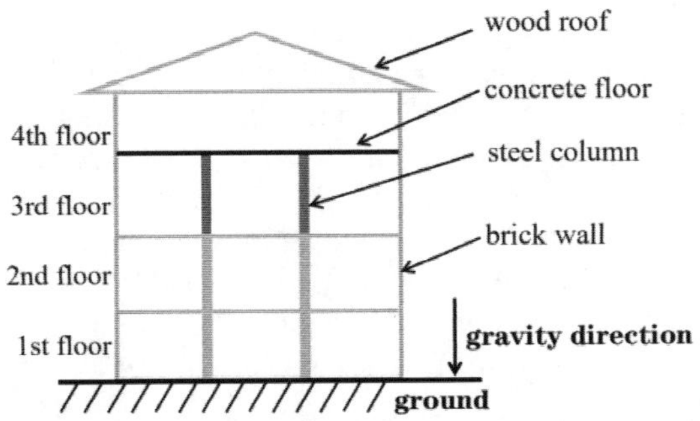


**Figure 3.** Schematic diagram of building structure

The project is a model project for educational use after the renovation of the existing historical buildings in the central urban area. In order to meet the requirements of large space, multi-pattern and flexible architectural function layout of innovative education, the building adopts the construction technology of "retaining the exterior wall as a whole and replacing the internal structure" to transform the existing buildings as a whole, just like "replacing the tank of the hot water bottle", this "replacing tank" method is suitable for upgrading the interior space of historical buildings [1,2], so as to achieve the purpose of activating and renewing historical buildings in the central urban area. It belongs to the renovation of old buildings under the conditions of "making track in snail shell" in typical central urban area.

### 2.2. Space optimization

The interior space of the building is designed according to the modern educational concept to highlight the use of public space and provide a reasonable space layout for stimulating students' innovation [3]. Based on the new teaching mode of innovation, openness, sharing and interaction, a green line with sustainable development is designed through the entire space of the building, and the diversified space with free, individual and flexible is designed around the green line

*2.2.1 Diversified Building Green Line.* The building adopts a variety of sustainable technologies. For example, the roof with solar panels can provide part of the indoor electricity; collecting roof rainwater can be of secondary use, such as watering the indoor plants, configuring "urban agriculture ", and with the indoor and outdoor environment to form an ecological cycle system. All of these devices are of open innovation platform for innovative course teaching under the premise of ensuring their use function. Students can not only learn and understand its working principle intuitively, but also form teams in related fields to carry out special research and complete in-depth development. All these energy-saving technologies, ecological projects, sustainable courses form a green line that runs through buildings ( as shown in Figure 4). The diversity of space in the green line will create a innovative space full of inspiration for students. The specialization of equipment and solutions in the green line will provide support for "problem-oriented" teaching methods, and implant the frontier problems in the real world into the students' daily study life.

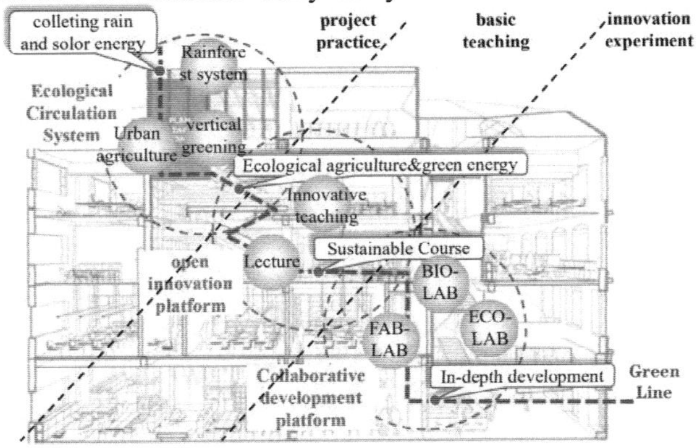


**Figure 4.** Green line and architectural space layout

*2.2.2 Functional spatial allocation.* All kinds of teaching space are allocated around the Green Line, making it of freedom, individuality, flexibility and adaptability, and highly meet the requirements of versatility and diversity as PBL(Project-Based Learning) teaching methods need. All allocation design is "student-centered", fuzzy functional partition, fuzzy space separation, dramatic integration of functional needs, interesting experience and innovation interaction, while providing the participation opportunities for teachers, students and users of the innovation field. Many international innovation laboratory, such as FABLAB(Fabrication Laboratory), LIVINGLAB(Living Laboratory) and BIOLAB(Biology Laboratory), are also distributed inside the building( as shown in Figure 3). It not only serves innovative teaching, but also becomes the engine of continuous innovation in campus space.

*2.2.3 Flexible spatial combinations.* In order to comply with Tongji Huangpu Design Creative Middle School to emphasize the happy, relaxed and multi-functional teaching environment, nine kinds of classrooms were specially designed (as shown in figure 5), and the expansibility was studied. Increasing the combination forms and function arrangements will be beneficial to ensure the space use efficiency of building at No.95 Huqiu Road. In the future, in the operation of Tongji Huangpu Design Creative Middle School, the space of each classroom could be flexibly adjusted according to the teaching contents, so as to achieve the best effect of courses.

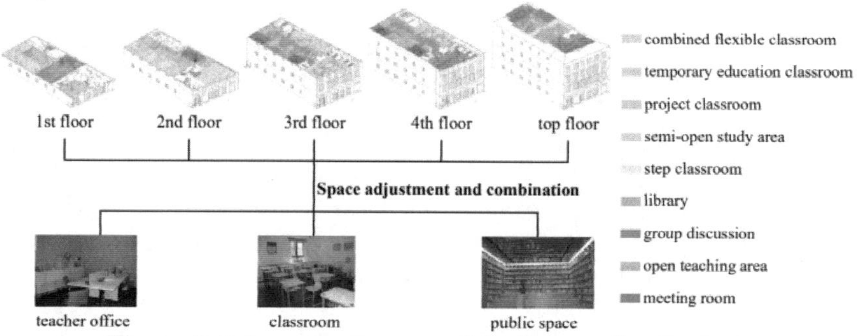


**Figure 5.** Schematic diagram of the 9 types distribution

## 3 Structural repairs

*3.1. Structural condition before repair*

The selection and arrangement of components are unreasonable and there are weak points. There are many damages existed in the building, such as cracks in wall and brick columns, paint drop, seepage of roof and wall, mild mildew, split of beams, steel corrosion and so on. Especially, the cracks of walls and brick columns in second floor and the splitting damage of individual beams have become dangerous points. The comprehensive seismic capacity of the building does not meet the requirements. The seismic-bearing capacity of the wall from first to third floor does not meet the requirements, while the seismic-bearing capacity of the wall in fourth floor meets the requirements. The slope of the southeast corner of the building is large, and its maximum slope is 15.1‰ eastward, which exceeds the limit value [4,5].

*3.2. Repair recommendations*

In order to improve the seismic capacity of the building as a whole, it is suggested that the originally support parts, such as internal wooden columns, steel columns, wooden floors, wooden roof, should be removed, retain the exterior wall of the building, and re-install the load-bearing system inside. The new support foundation should be removed from the existing foundation. It is also suggested that a ring beam structure should be added to the building wall to improve the integrity of the building. Reinforcement and maintenance measures should be adopted immediately for the components that have forms the danger point. For the damaged walls and brick columns in the second floor, the pressure grouting should be carried out first, and then the reinforcement treatment should be carried out by using the method of reinforced mortar surface. For the structural cracks in brick wall, it is suggested to use pressure grouting for sealing treatment. For the serious slope at the southeast corner of the building, it is suggested that the external reinforced mortar surface or other effective methods should be used to reinforce the walls. For other damages, it is suggested that repairs be carried out in accordance with the relevant provisions of the Technical Regulations for Building Repair Works (DG/TJ08-207-2008) [6].

### 3.3. Structural repair programme

Using the method of "remaining the entire external wall, replacing the entire internal structure" to upgrade and renovate the existing building just like "replacing the tank of the hot water bottle", as shown in figure 6. The difficulties in this repair process is: relative relationship between old and new foundations, structure design of internal steel system, reinforcement measures of the external wall, connection forms of external walls to new structures.

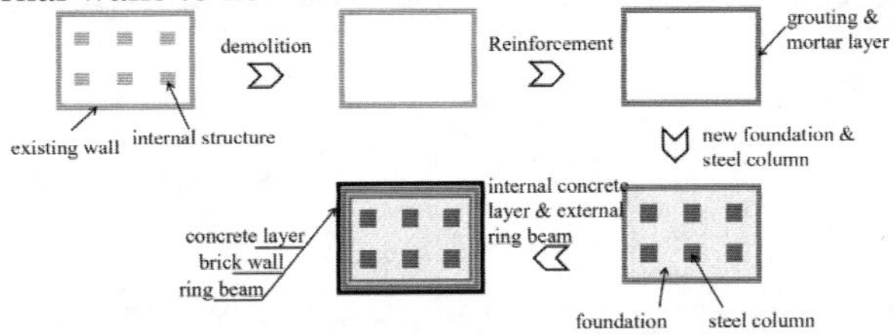


**Figure 6.** Schematic diagram of structural repair plan

1) Relative relationship between old and new foundations. The support piles of the new foundation adopt the type of hollow pipe anchor, with size of 250×250 mm, and with length of 22 m. The number of support piles is 166, and the characteristic value (design value) of pressure-bearing capacity of each pile is 245 kN, all the piles is support on the gray silty clay. The new and existing foundation are separated, and their relative relationships are shown in Figure 7.

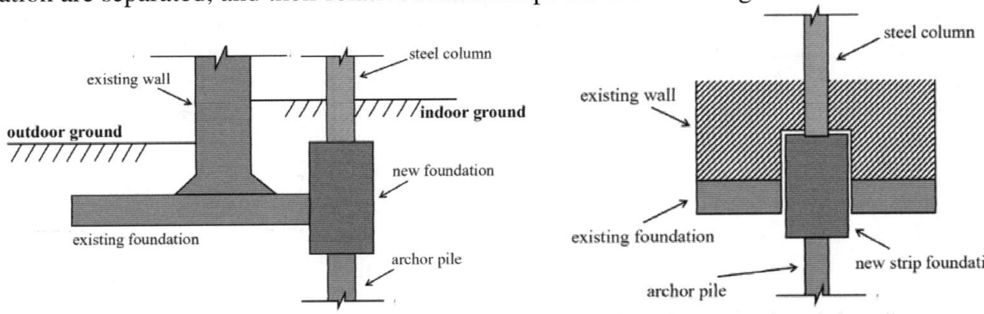


(a) Relative position inside the building (b) Relative position when passing through the wall

**Figure 7.** Relative position between new and existing foundation

2) structure design of internal steel system. The internal steel system is supported on the new foundation and is reliably connected to the retaining exterior wall. The overall stability of the steel structure meets the requirements of the specification by setting inclined braces in the appropriate position

3) reinforcement measures of the external wall. During the process of internal demolition, the external wall has lost horizontal restraint, and before any stabilization measures, the calculated length of the exterior wall is increased from the original height of the floor to the height of the whole building. The external wall changes from load-bearing to non-load-bearing wall.

Temporary horizontal support is added to the external wall in the process of demolition for better stability. At the same time, the outside of the existing external wall is added ring beam, and the inner side is added a layer of reinforced concrete slab to improve the stiffness and overall stability of the external wall; In addition, the outer ring beam is connected with the inner concrete slab through the pin key to form a overall structure, as shown in figure 7.

4) connection forms of external walls to new structures. When the internal steel structure is completed, the vertical load is mainly transmitted to this steel structure and nearly not transmitted to the existing external wall in principle, but the external wall is still used as the peripheral structure of

the school building, and the reliable connection between the external wall and the steel structure is needed. After the internal steel structure is completed, the concrete slab is firmly connected with the steel structure through the embedded anchor plate in the concrete slab, as shown in figure 8.

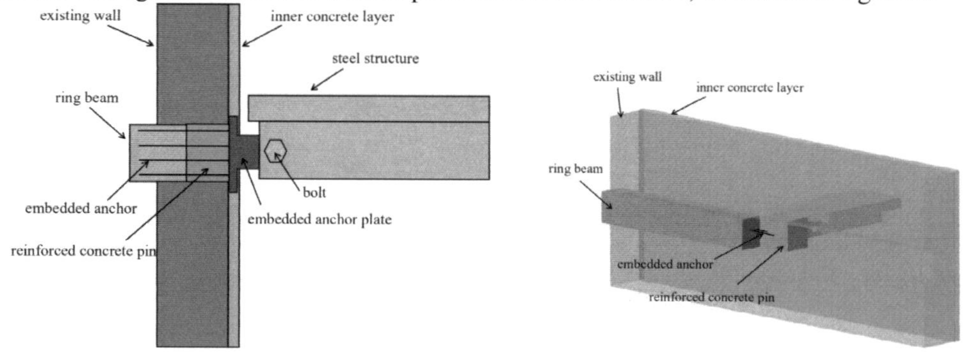


(a) 2D schematic (b) 3D schematic

**Figure 8.** Connection between external wall and steel structure

## 4 Organization and management

Difficulties: 1) Arrangements for demolition works; 2) Transporting and Stacking of materials and garbage; 3) Schedule arrangement for severe weather and major events; 4) Construction of the internal pile foundation; 5) Design and implementation of fireproof equipment

### *4.1 Demolition process*

Improper demolition process will inevitably destroy the integrity of the school building, greatly reduce the stability, resulting in the serious crack problem. In this project, the reasonable demolition sequence is adopted in the demolition process. First, the old floor inside was removed; and the temporary internal supports was made. Then the wooden beam, column, part of the load-bearing brick wall, and the log truss roof were removed in sections. After this, the whole external wall was strengthened to maintain stability, and then the new internal steel structure was made and collected together.

### *4.2. Stacking and transporting of materials and garbage*

East side of the project toward Huqiu Road, and the other three sides are adjacent to public and residential buildings, the construction site is very small. So it's a challenge to get rid of the garbage and move in the needed materials. All the garbage is stacked inside the building or at the north road which width is less than 5m, and then transported at night without affecting the normal living conditions of the surrounding public. At the same time, materials are also carried in at night, the roof position, which are removed at first, will be used to move in the materials by crane. Choosing the appropriate stacking position and transportation time solves the difficult problems of how to move numerous garbage and materials.

### *4.3. Schedule arrangement for severe weather and major events*

The construction period encountered typhoon, heavy rain and other bad weather conditions for many times, as well as traffic control during the Expo, all of these will inevitably effect the construction schedule. Except having to stop work, most processes that are insensitive to external conditions are carried out as normal as possible to reduce the impact of weather on the duration of the project.

### *4.4. Construction of the internal pile foundation*

In order to reduce the influence of vibration, soil-extrusion and excess pore water pressure on the surrounding environment during the construction of pile foundation, taking the following protective measures: 1) Layout of pressure release holes in adjacent roads and buildings; 2) The piling sequence deviates from the protective object and reasonably controls the pilling rate. 3) Monitoring the surrounding environment during piling, Information construction ensures the safety of surrounding buildings and pipelines, the maximum monitoring deformation not exceed 9 mm, less than the allowable value.

*4.5. Renovation of fire facilities*
In addition to implementing the mandatory provisions of fire protection requirements, new indoor fire hydrant system is adopted to enhance fire protection capability.

The system adopts the form of temporary high pressure water, connecting two diversions from the municipal water network, and the outdoor water pipeline uses the municipal pipeline to form the outdoor network together, and the valve is added to the municipal pipeline between the two channels of water intake.

In addition, two fire hydrants(one for use and another for backup) and fire hydrant pumps (one for use and another for backup) were set near the water pump room,. Air pressure tank was installed as required, and one pump connector was installed outside.

## 5 Conclusions
At present, it is imperative to upgrade the function of school buildings in high density area. Aiming at the dual task of protecting historical buildings and adapting to the teaching requirements of innovative education in the new era, reasonable measures have been taken in the design and construction, which lasted nearly a year and a half, and the school building renovation of 2500 $m^2$ at No.95 Huqiu Road has been completed.

In space allocation, the teach office area is reduced while the public space is expanded according to the training requirements of innovative talents. a green line that integrates teaching, practice and innovation go through the entire building. A free, personality and flexible space is set up around this green line to adapt to the project-oriented multi-function and diversity space.

In structural design, the "replacing tank" method is applied to realize the upgrading and renovating of school building function. The internal new foundation adopts reinforced concrete structure composed of anchor piles and its upper strip concrete. This new foundation is reasonably separated and traversed from the existing external wall foundation. The internal new steel frame is reliable connected to this new foundation and the existing exterior wall. In addition, a reinforced concrete layer is added to the inner surface of the existing wall to increase the strength and overall stability.

Technological innovation plays an important role in the construction and management of engineering projects. The implementation of the project has overcome many difficulties, such as narrow construction site, material transportation difficulty, hoisting difficulties, adjacent to the residential buildings, dust and noise, controlling the settlement of surrounding historical buildings and so on. The internal steel structure adopts the prefabricated installation, this kind of assembly construction have accelerated the construction progress. Reinforced concrete layer has be add to the inner side of the existing wall to reinforce the entire building. The original fire facilities have be overall upgraded, and the ground settlement was automatically monitored, all these construction works were in ordered operation and efficient process.

## 6 References
[1] Yan Lihui. A Brief Analysis of"Thermos Bottle Liner Change" Renovation Mode of Historical Building --A Case Study of the Renovation of No.179 Block East Nanjing Road[J]. Chinese & Overseas Architecture, 2018(8):138-140.
[2] Zhang Jianming. Quality Control of Old Walls during the "Hot Bottle Replacement" in Historical Buildings[J]. Development Guide to Building Materials, 2018(8):138-140.
[3] Liu Xin. Innovative Application of "Open" Teaching Mode in "Living Space Design" Course[J]. New Curriculum Research, 2018(12):68-69.
[4] Tongji University Building Quality Inspection Institution. Damage detection report of No. 95 Huqiu Road, Huangpu District, Shanghai[R]. 2017,1.
[5] Shanghai Municipal Housing Design Institution Co.,LTD. Earthquake Resistance Evaluation Report of No. 95 Huqiu Road, Huangpu District, Shanghai[R]. 2016.9.

[6] Shanghai Housing, Land and Resources Administration. Shanghai Construction Engineering Code <Technical Regulations of Housing Repair Engineering> (DG/TJ08-207-2008)[S]. Shanghai, 2008.

# Building Renovation Process Map in Private and Public Works to Improve Communication and Data Exchange

**Martina Signorini[1,*] and Sonia Lupica Spagnolo[1]**

[1]Politecnico di Milano, Department of Architecture, Built environment and Construction engineering, Via G. Ponzio 31, 20133 Milano, Italy

[*] martina.signorini@polimi.it

**Abstract.** Along the building process the cooperation among several stakeholders and the integration of information during the whole project life cycle, whether it is a new project or a renovation intervention, are two key essential factors. In fact, in AEC sector the loss of information, the lack of data and the improper or repetitive exchange of information are common issues that may cause delays and cost increase. To enhance communication and data transfer, an appropriate rationalization of information flows is necessary. Starting from the outcomes of BIM4EEB, an ongoing Horizon2020 project, the paper analyses the workflow processes to understand how to optimise information workflows among different actors and along different stages by the use of ICT and BIM-based tools. Particular attention was payed to individualise the differences between public and private works in order to provide methods and apply the BIM4EEB approach to renovation process in both cases.

## 1. Introduction

Among the objectives of 2030 Agenda, there are a 40% reduction in GHG emissions, a rate of 32% for renewable energy and at least an increase of energy efficiency by 32,5% [1]. According to a BPIE analysis on EPC data, about 97% of building stock, currently not in A level, must be upgraded to achieve 2030 decarbonisation objectives [2]. If we consider that buildings are the biggest energy consumers – they consume about 40% of energy and they are responsible for 36% of $CO_2$ [3] – hence they are strategic to reach the set of environmental goals. Furthermore, the condition of the environment where a person lives is very important if we think that generally people spent 90% of time indoors and that indoor air quality ($CO_2$, PM2.5, PM10, VOC) impacts on our health [4]. Consequently, renovating the building stock means also to give a better life quality to inhabitants.

Ideally, buildings are subjected to renovation and refurbishment during their lifecycle according to a specific schedule: the owner should guarantee proper and good building conditions thanks to professionals' surveys, but also to planned and executed maintenance activities. However, the renovation process is slowed down by other factors such as: not understanding a long-term plan and its importance; no interest in thinking about the building in long term but preference to have short-term savings; buildings technical complexity due to their age and state of deterioration; the non- or semi-professional ownership of the residential buildings; disturbing inhabitants by noise caused by renovation or by leaving their apartments; the alternating of several and different actors during the whole renovation process (from maintenance professionals to managers). Then, renovation needs a precise procedure to take into account contemporarily several aspects such as costs of the renovation interventions, performances to be reached (e.g. indoor thermal comfort levels, $CO_2$ emissions, energy, safety), national building regulations that could vary among different Countries, building specific information and end user's needs.

Within this context, the BIM4EEB project, an ongoing Horizon2020 project founded by European Union, wants to face these issues by producing a BIM model that is enriched of fundamental information of the building and that could be considered the basis for the cost-efficient renovation planning, implementation and future renovation phases [5][6].

A specific task of the project deeply analysed the renovation process focusing on several information workflows for the directly or indirectly involved actors and outlining a complete process map in case of renovation. Designers, construction companies, facility managers, energy service companies, owners and inhabitants have then been assumed as main users of the BIM management system, highlighting which activity they do and point out their specific information needs and requirements. Attention has been paid also to when the activity is performed focusing on the stages of the building process. Finally, an integration of these single elements was made, showing up the differences in activities and actors if public or private works were taken into consideration.

## 2. The renovation process map development

BIM4EEB serves the purpose of developing and testing in real environment a BIM toolkit for renovation interventions of existing residential buildings in Europe. As mentioned, part of the initial effort was intended to analyse each step of the renovation building process with the focus on information workflow for each actor in order to make it replicable using the BIM management system, finding possible shortcuts in data exchange but also understanding how the building process could become more efficient by exploiting the BIM4EEB toolset.
The study is characterised by six steps:

1. Individualizing the involved stakeholders in the renovation process. The experience of Consortium partners and a literature review based on the analysis of California Commissioning Collaborative, Global Buildings Performance Network and RIBA works lead to a list of the main stakeholders with their relative definition [7][8][9];

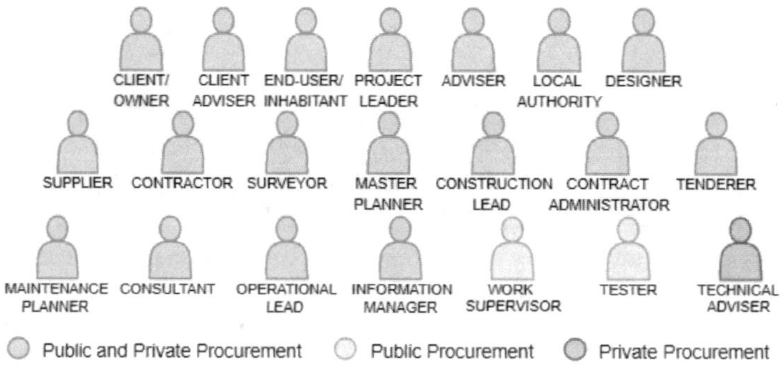


**Figure 1.** Main stakeholders involved in renovation process, considering the division between public and private works.

2. Outlining the undertaken activities;
3. Describing the outputs of each specific activity;
4. Outlining the stages when the activities occur according to the structure of EN 16310: 2013 [10];

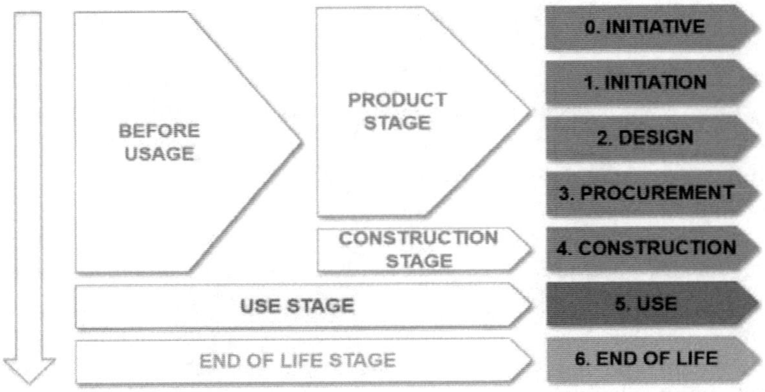


**Figure 2.** Stages in the building asset lifecycle according to EN 16310.

5. Finding out possible differences among different Countries;
6. Identifying possible optimisation alternatives.

To collect these data, two spreadsheets were created, one for public works and the other one for private works. These analytical spreadsheets have been later converted into a comprehensive flowchart, a process map that takes into account both of them.

## 3. Process analysis in private and public works

The renovation process has some distinctions between private and public works. In case of private works the owner is a private person, a group of people, a private company or some private companies. In public works, the owner is a public administration.

Activities are very similar, but sometimes objectives and procedures change, the subject of the action also could be impersonated by a different actor (Figure 3, Figure 4).

### *3.1. Private works and public works: main differences*

Private works are generally profit-oriented and the client/owner leads the works with the final goal to obtain an economic benefit and he/she decides whether to undertake certain work after having made analysis about cost-benefit and available funds. Among these objectives there is also the improvement of building performances and the enhancement of inhabitants/end-users' conditions and comfort. Public works are managed by governmental bodies/authorities and they focus on public interest, trying to provide social and economic benefits. The improvement of building performance and the enhancement of inhabitants/end-users' conditions and comfort are objectives taken into consideration too.

Within the private works context, the budget is provided by the client/owner through personal private funds or banks and/or private investor groups, also named party financier. In detail, in the so-called initiation phase, the Client Adviser prepares a detailed business case that can be used for finding financing from third parties, and later the bank or the third party financier assesses it and the feasibility of project by own cost assessments and audits from external consultants to determine if the investment is safe. If the assessment is successful and the bank/third party financier agrees to finance the project, the next step for the client/owner is to negotiating financing terms (e.g. percentage of project financed, how money will be delivered over the project, interest rate, time frame and payments schedule, etc). Finally, both parties sign the financing contract. The funds in the case of public works arise from public funds such as tax revenue, duties, and fees. The client/owner must approve the preliminary design and spending plan before moving to the design phase.

Another significant moment is characterized by private bidding. The project leader undertakes the tender action converting the project design into a set of pre-contract documents and a set of tender requirements. The client/owner can decide to choose one contractor rather than another one according to the fulfilment of the previous requirements and not necessarily on the basis of the best economic offer. On the other hand, in public sector the bidding is a more competitive process and characterised by more transparency, open to the public. The contract is awarded to the lowest bidder to guarantee the responsible use of money.

The procurement strategy selection is another point that public and private works face differently. Private works generally have a faster procurement process since there is not publicly subsidised, in addition the Project Leader can chose the supplier/subcontractors he/she prefers as long as they are in the budget. Similarly, private projects may, where required, procure the necessary materials and services.

In the case of public works, however, this procedure is not so rapid because it is constrained by local, regional, national or international legislations. Procurement practices must be approved by governmental bodies, suppliers/subcontractors can carry out inspections in order to qualify to bid and in this way the process is slowed down. "European Union Procurement", with the purpose to open up public procurement within the EU, guaranteeing the free movement of supplies, works and service, defines and established the public procurement rules that are applied to any public purchases.
Although the directive "European Union Procurement" defines public procurement rules, there are differences among countries. Therefore, especially in renovation interventions, the workflow process could vary according to the Country.

Considering actors in private and public works, a limited number of people are involved in private works, with owner and project leader having key roles. Instead public works have a major number of users since the number of activities is higher too.

| WHO | MAKES | WHAT | WHEN | SHORTCUT |
|---|---|---|---|---|
| Project leader | Defines | the project organizational structure and range of consultants and others to be engaged forthe project, including definition of responsibilities | Initiation | |
| Project leader | Defines | possible procurement strategy | Initiation | |

**Figure 3.** Extract of a renovation workflow: activities of a renovation process in case of private works – Initiation phase.

| WHO | MAKES | WHAT | WHEN | SHORTCUT |
|---|---|---|---|---|
| Project leader | Defines | the project organizational structure and range of consultants and others to be engaged forthe project, including definition of responsibilities | Initiation | |
| Project leader | Evaluates | if it is necessary to appoint an external professional | Initiation | If the owner has an employee with the expertise to be appointed as site surveyor |
| Client/Owner | Appoints | designers to be engaged for the project (such as structural, mechanical, electrical, HVAC, geotechnical, fire security, acoustics, lighting, etc.) | Initiation | |
| Project leader | Defines | possible procurement strategy | Initiation | |

**Figure 4.** Extract of a renovation workflow: activities of a renovation process in case of public works – Initiation phase.

### *3.2. Public works: regulations*

BIM4EEB project involves partners from different Countries (Italy, Finland, Spain, Sweden, Ireland, Cyprus, Germany, Belgium, and Poland) and the BIM toolkit will be developed and then applied to three residential buildings respectively in Italy, Poland and Finland. Due to this fact, national differences in the procurement regulations were analysed.
The public works process is ruled by specific European directives that are transposed into local national laws:

- Directive 2014/24/EU on public procurement [11];
- Directive 2014/25/EU on procurement by entities operating in the water, energy, transport and postal services sectors [12];
- Directive 2014/23/EU on the award of concession contracts [13].

The main principles of these directives are transparency, equal treatment, open competition and the possibility for EU companies to access to fast and efficient review. For putting public funds to good use, it is fundamental to reach a competitive, open, and well-regulated procurement market [14].
Concerning Italy, the regulation for public works is contained within the document "Italian public contract code - Legislative decree No. 50, April 18, 2016" [15] that outlines rules about all phases of the project from the planning to the award, also covering testing and verification of compliance and relates to other specific regulations (e.g. Italian Code for the construction design and Legislative Decree No. 42 of 22 January 2004 which contains the Code of Cultural Heritage and Landscape).

In Poland public works follow the "Public Procurement law": it defines the rules and procedures for awarding public contracts, legal protection measures, control of the award and the competent authorities [16]. Even "Construction Law" and "The technical conditions which should be met by buildings and their location" are considered in regard for public and private works in this Country [17].
Finnish legislation on public procurement is instead governed by the "Act on Public procurement and concession contracts", the "Act on procurement and concession contracts of entities operating in the water, energy supply, transport and postal services sector", and the "Public defence and security procurement act" [18][19][20]. Other areas of national laws relevant to public procurement include the Act of the openness of government activities and the Act on the Contractor's Obligations and Liability when work is contracted out [21][22]. The first document refers to the phase when tender offers and other documents about public procurement are to enter the public domain and the extent of parties' right to access to information; the second one instead requires, when work is contracted out, that the party responsible for the project have guarantees that subcontractors will fulfil their various obligations.

## 4. Conclusion

The inefficiencies due to incorrect or redundant information are typical issues in construction sector, hence the need to precisely track the process map and find shortcuts.

The paper has been developed in the context of the EU founded project BIM4EEB. It proposes the methodology adopted to analyse and optimize the information workflow within a renovation process outlining the differences between private and public works.

Among the main differences, there are: from one hand profit-minded concept in private works and on the other public concerns and social, economic and/or aesthetic benefits in public works; different sources of funding and different criteria for selection of procurement strategies; regulation limitations to be followed in public works.

Starting from these activities, it is possible to map their outputs in a traditional process and in BIM-based processes and to assess when each activity is undertaken and its expected outputs. Finally, to define how ICT and BIM can support the rationalisation of information flows, a synoptic flowchart that shows the complete process map has been developed.

**Acknowledgments**
The BIM4EEB project has received funding from European Union's H2020 research and innovation programme under grant agreement N. 820660. The content of this publication reflects the authors view

IOP Conf. Series: Materials Science and Engineering 1044 (2021) 012012 doi:10.1088/1757-899X/1044/1/012012

only and the Commission is not responsible for any use that may be made of the information it contains.

## 5. References

[1] European Commission 2014 2030 climate & energy framework [Online] Available: https://ec.europa.eu/clima/policies/strategies/2030_en
[2] Buildings Performance Institute Europe (BPIE) 2017 97% of buildings in the EU need to be upgraded
[3] European Commission – Department: Energy 2020 Energy efficiency in buildings
[4] World Health Organization (WHO) 2014 Combined or multiple exposure to health stressors in indoor built environments - An evidence-based review prepared for the WHO training workshop 'Multiple environmental exposures and risks'
[5] BIM4EEB 2019 BIM4EEB: BIM based toolkit for Efficient rEnovation in Buildings [Online] Available: https://www.bim4eeb-project.eu/
[6] BIM4EEB 2019 D2.1 Definition of relevant activities and involved stakeholders in actual and efficient renovation processes
[7] California Commissioning Collaborative 2006 California Commissioning Guide: Existing Buildings [Online] Available: https://www.cacx.org/resources/commissioning- guides.html
[8] GBPN 2013 What is a deep renovation definition? [Online] Available: http://www.gbpn.org/reports/what-deep-renovation-definition
[9] RIBA 2013 Plan of Work 2013 – Overview [Online] Available: https://www.ribaplanofwork.com/
[10] European Standardization Organizations 2013 EN 16310:2013 - Engineering services - Terminology to describe engineering services for buildings, infrastructure and industrial facilities
[11] European Union 2014 Directive 2014/24/EU of The European Parliament and of The Council of 26 February 2014 on public procurement and repealing Directive 2004/18/EC *Off. J. Eur. Union* pp 65–242
[12] European Union 2014 Directive 2014/25/EU of the European Parliament and of the Council of 26 February 2014 on procurement by entities operating in the water, energy, transport and postal services sectors and repealing Directive 2004/17/EC *Off. J. Eur. Union* pp 243–374
[13] European Union 2014 Directive 2014/23/EU of the European Parliament and of the Council of 26 February 2014 on the award of concession contracts *Off. J. Eur. Union* pp 1–64
[14] European Commission 2020 Public Procurement [Online] Available: https://ec.europa.eu/growth/single-market/public-procurement_en
[15] President of the Republic 2016 Legislative decree No. 50, April 18, 2016
[16] Public Procurement Office 2004 The Act of 29 January 2004- Public Procurement Law
[17] Minister of infrastructure 2017 Ordinance of the Minister of Infrastructure of November 14, 2017 on the technical conditions to be met by buildings and their location
[18] Finland Ministry of Economic Affairs and Employment 2016 Act on Public Procurement and Concession Contracts
[19] Finland Ministry of Economic Affairs and Employment 2016 Act on public contracts in special sectors
[20] Finland Ministry of Defence 2016 Act on Public Procurement in the fields of defence and security pp 1–52
[21] Finland Ministry of Justice 1999 Act on the Openess of Government Activitie*s* pp 1–16
[22] Finland Ministry of Economic Affairs and Employment 2015 Act on the Contractor's Obligations and Liability when Work is Contracted Out

IOP Conf. Series: Materials Science and Engineering 1044 (2021) 012013 doi:10.1088/1757-899X/1044/1/012013

# Define the End Users Preferences and Issues among Residential Houses in the UAE: Citizens vs. Non-Citizens

**M Albattah*, E AbuNamous, S Mansour, and L Bande**

United Arab Emirates University, Al Ain, United Arab Emirates

*mohammed.battah@uaeu.ac.ae

**Abstract.** One of the most desirable things for human beings is to own a house. Some governments, such as in the United Arab Emirates (UAE), have a system with different programs for their citizens to own the house. Generally and by applying one of the available programs, there are two ways to own a house in the UAE. The first way is to get a loan and buy a house that is already built (*constructed house*), and the second way is to get a loan and build a house on owned land (*non-constructed house*). The main purpose of this study is to define the citizens' preferences between these two options, furthermore the issue based on these preferences and compare them with the non-citizens. Also, this study explores their perceptions of number of features that can be applied in their future house. The authors applied a Simi-structured interview targeting the current and potential house owners in Al-Ain city, UAE. The authors found that most of the participants prefer to build their own (*non-constructed house*) instead of buying a (*constructed*) one. Also, the most desirable features that the participants prefer are the ones they can benefit while occupying the house (i.e. flexibility to add or remove spaces, and flexibility to change the house elements).

## 1. Introduction

The housing industry plays a major role in community stability for any country. Moreover, houses are considered as the lowest and first level of needs in the Maslow's hierarchy of human needs that is reflected an important role in human life [1, 2]. Therefore, any government should consider the importance of housing, and make sure that supply meets demand by providing financial institutions, or by providing affordable and adequate housing [2, 3].

The intention of the UAE government is to help citizens own their houses by two different options either by conducting several public housing projects, *constructed houses*, or by giving a loan and build houses, *non-constructed houses*, in different parts of the country. In recent years, the UAE citizens have started to use new designs and new technologies in their houses to be more suitable for their culture and satisfy their needs related to safety, health, security and quality [4]. However, there are a number of factors that affect the demand and supply of the housing industry that governments should consider. These factors include population growth, household growth, unequal distribution of dwelling units, and end user unemployment rate and their income level [2, 5, 6]. However, there are limited research about the end users preferences on the residential houses.

The aim of this study is to define the citizens' preferences and issues among the two ways of owing a house, *constructed house, and non-constructed house*, and compare them with the non-citizens, focusing on Al-Ain city. Also, it will explore their perceptions among number of features that can be applied in their future house. These features will help to suggest a new and different construction method for the houses that will contribute to meet the UAE's Vision 2030.

## 2. Literature review

The housing sector faces two main problems; the quantitative and qualitative problems. The quantitative problems relate to the provision (state or private) of houses that the government provide, which does not meet the housing demand. Whereas the qualitative problems relate to the type of the houses that does not meet the end user comfort, social, culture and religious needs. Both quantitative and qualitative problems affect the quality of their life in different ways and of course the quality will affect the satisfaction of the end user. However, there are two approaches to assess the quality of the building, which are objective and subjective approaches. The objective approach includes the evaluation of the physical characteristics, facilities, services and environment, which are widely used [7, 8]. However the subjective approach includes the evaluation perception, satisfaction and aspiration. In general, to improve the quality of houses, the government should understand the end user needs and their satisfaction. It is an essential method that helps planners, architects, developers, and policymakers to predict the future needs and requirements of the end user. Providing more houses and without considering their quality will not solve the housing sector problems [8, 9, 10].

Several studies shows that there are some factors that affect the end user satisfaction. One of these factors is related to the dwelling unit design, such as the layout of the floor plan and the distributions of the internal spaces like the living, dining, and bedrooms. Some of the end users do not prefer to buy a *constructed house* because of this factor. The floor plans and the layout of the *constructed house* does not meet their need and desire [8, 11, 12]. The other factor is related to the location of the dwelling unit and the nearby services, such as public facilities, social facilities, and environment and neighbourhood facilities, which also affect the end users decision to buy a *constructed houses* in areas that did not have the required services [8, 10]. However, when the end users decide to build the house in their own land *non-constructed house*, they will face other factors that are related to the construction industry [12].

## 3. Methodology

This study is applied a qualitative research method and collected data regarding the end users' perception. The authors conducted semi-structured interviews, in the time between February 2019 and July 2019, focused in Al-Ain city. Three hundred and forty (340) participants were included in this study, with different demographic information. A comparison between citizens (30% of the total participants' number) and non-citizens (70% of the total participants' number) was applied in this study to find any differences among their perceptions. The authors asked the participants about their preferences regarding the house type *constructed house vs. non-constructed house* and the reasons behind their choice. Also authors asked the participants about their perception on a number of features that can be applied in their future non-constructed house (i.e. exact cost with clear and detailed design, fast execution within a few weeks, relocation of the house, flexibility to quickly change the elements of the house such as the walls, doors, and windows, also flexibility to quickly add spaces or remove spaces, and involvement with one party instead of number of parties before, during, and after execution). The authors used SPSS statistical software to conduct the results.

## 4. Results

There are two parts in this section. The first part includes the participants' preferences regarding the house type and their reasons behind it, and the second part includes the participants' perceptions on a number of features that can be applied in their future house.

### *4.1 House type preferences (constructed vs. non-constructed)*

The authors asked the participants if they already own their house, and if they planning to own a house in the future. Around 28% of the citizens and non-citizens own their houses. However, 95% of the citizens are planning to buy or construct a new house in the future, while 90% of the non-citizens are planning to buy or construct a new house in the future. The reasons behind that is because some of participants do not own a house yet; some of them want this future house for their children, some of them want this future house for the second wife, and some of them want this future house for investment purposes. The authors believe that the remaining participants were satisfied with their current house or they are not ready to buy or construct a house because they are still live with their parents or because they do not have enough money.

Furthermore, the authors asked the participants about the house type, specifically if the preferred a *constructed house* or (build their house on owned land) *non-constructed house*. The results show that 67.7% of the citizens and 58.1% of non-citizens preferred *non-constructed house*, which is the higher percentage (figure 1). However, there was no statistical difference between citizens (67.7%) and non-citizens (58.1%) in the rate of house type preferences, as demonstrated using chi-square test (p = 0.10). Therefore, citizens and non-citizens have similar desire and prefer to build their own house, *non-constructed house*, when they have the chance.

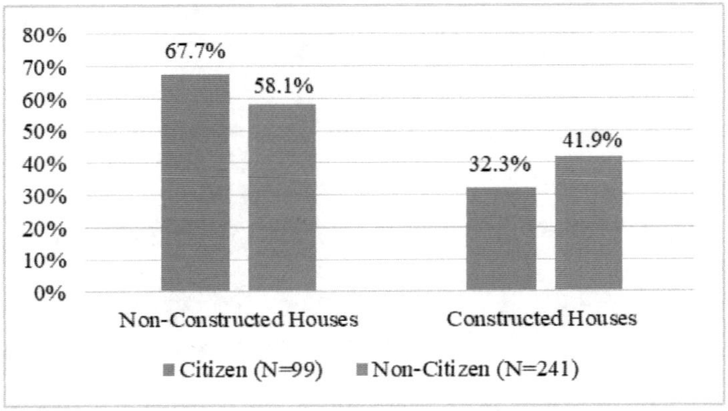


Chi-square test was performed to test the difference between the columns. The P-value = 0.10.

**Figure 1.** House Type Preferences (Citizens vs. Non-Citizens).

Furthermore, the authors asked the participants about their motivations and reasons for wanting to own a *constructed house* (figure 2). Some participants gave a number of reasons, while other participants gave a single reason. The highest motivation and reason for the citizens and non-citizens to buy a *constructed house* is its layout and appearance, with 48.5% and 39.8% respectively. However, the least motivation and reason for the citizens and non-citizens to buy a *constructed house* is for investment purposes, with 4.0% and 2.1% respectively.

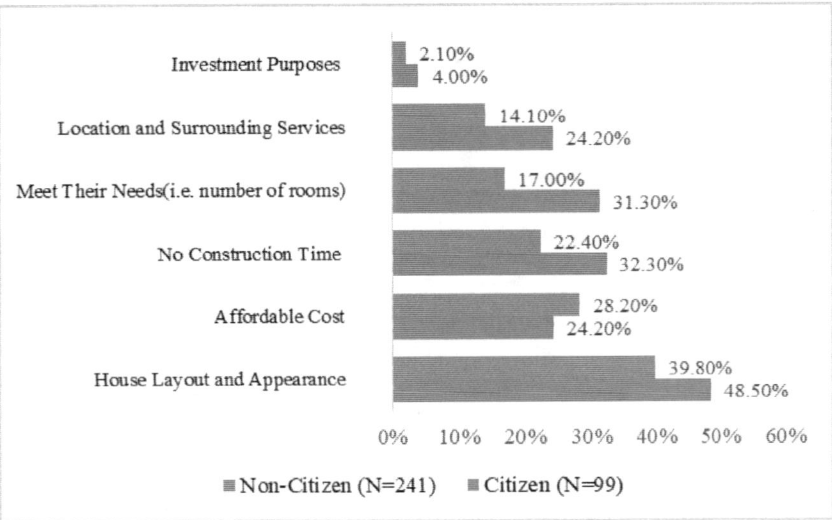


**Figure 2.** Motivations and Reasons for Owning a Constructed House (Citizens vs. Non-Citizens).

One the other hand, the participants explained their reasons for preferring *non-constructed house* instead of purchasing a *constructed* one (figure 3). The most reason for the citizens and non-citizens to build a *non-constructed house* is the fixable design that meet their needs (i.e. number of rooms), with 57.6% and 36.9% respectively. However, the least reason for the citizens to build their own house, *non-constructed house*, is guaranteeing the house quality during execution, with 12.1%, while the least preferred reason for the non-citizens to build their own house, *non-constructed house*, is the unique layout and appearance, with 5.0%.

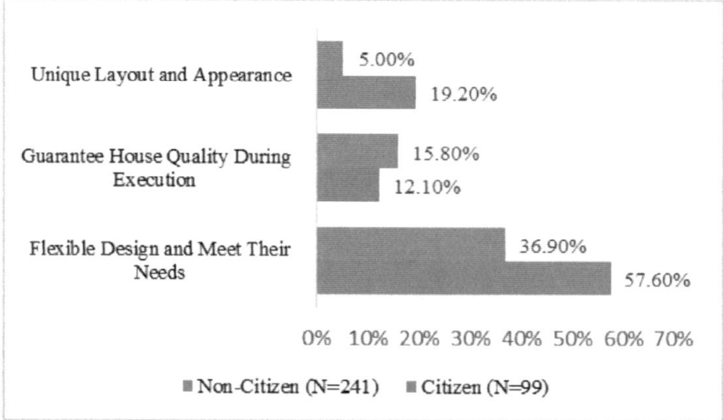


**Figure 3.** Motivations and Reasons for Owning a Non-Constructed House (Citizens vs. Non-Citizens).

*4.2 Participants' perceptions on the house features*

Furthermore, the authors asked the participants about their perception on number of features that can be applied in their future non-constructed house. They focused on the following features: 1) knowing the exact cost with clear and detailed design before execution; 2) fast execution within few weeks; 3) flexibility to relocate the house from one land to another land; 4) flexibility to quickly change the house elements like the walls, finishes, doors, and windows; 5) flexibility to quickly add spaces (i.e. extra rooms) or remove spaces; 6) involvement with one party instead of number of parties before, during, and after execution.

The authors explained these features to the participants, and asked them if they were interested in applying them in their future house. More than 50% of the participants (citizens and non-citizens) were interested in all the features (figure 4). The feature that the citizens were most interested in was the flexibility to add or remove spaces, with 93.9%, while the feature that the non-citizens were most interested in was the flexibility to quickly change the house elements, with 89.6%. However, the feature that the citizens were least interested in was the involvement with one party during execution, with 65.7%, while the feature that the non-citizens were least interested in was the fast execution within few weeks, with 58.5%.

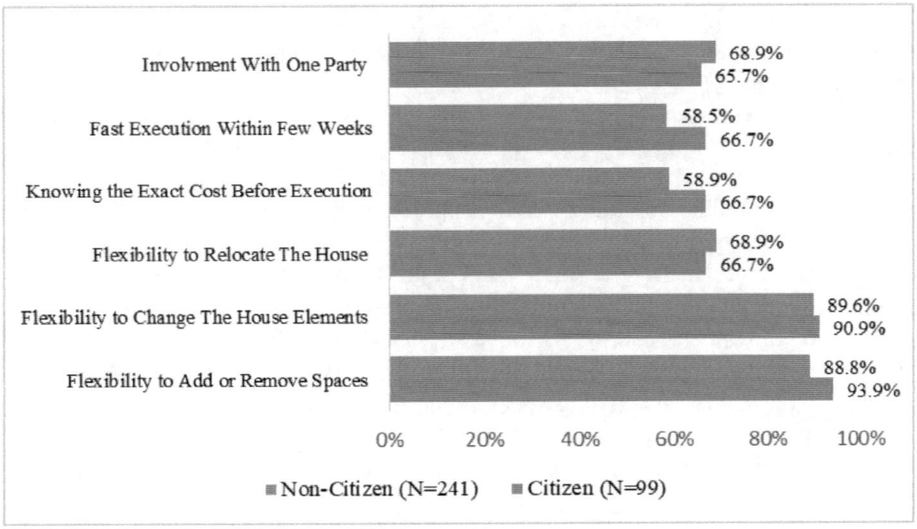


**Figure 4.** Perceptions on Features among Future Non-Constructed House (Citizens vs. Non-Citizens).

Looking to the results from different angle, the authors divided the features into three groups based on its time period of benefit. The first group is "the features before living in the house," which includes fast execution within few weeks, involvement with one party during execution, and knowing the exact cost before execution. The second group is "the features during living in the house," which includes flexibility to add or remove spaces, and flexibility to change the house elements. The last group is "the features after living in the house," which includes flexibility to relocate the house.

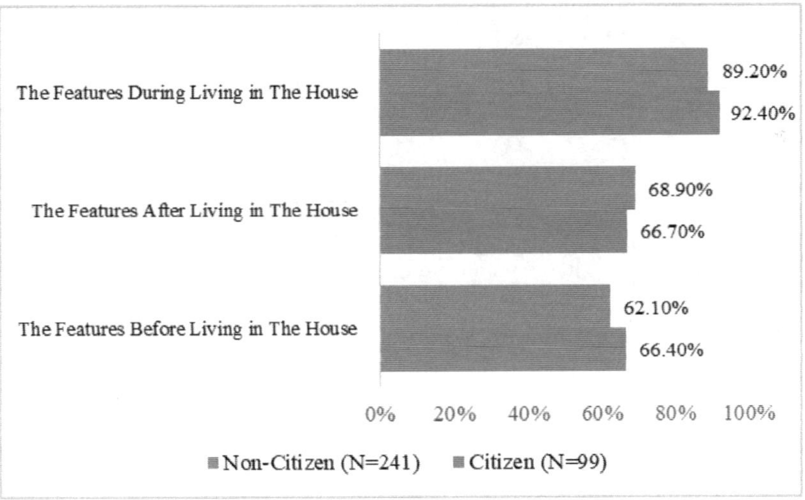


**Figure 5.** Groups of Features Based on Its Time Period of Benefit (Citizens vs. Non-Citizens).

The results clearly show that citizens and non-citizens prefer more the group of features that they benefit from during living in the house, with 92.4% and 89.2% respectively. However, the other two groups of features are almost the same. Although, citizens and non-citizens prefer less the group of features that they benefit from before living in the house, with 66.4% and 62.1% respectively (figure 5).

## 5. Conclusion

Using semi-structured interviews from February 2019 to July 2019, this study examined and compared the preferences regarding the house type *constructed house vs. non-constructed house* among citizens and non-citizens of the UAE, and the reasons behind their choice. Also, this study examined and compared perceptions regarding the number of features that can be applied in the future house among citizens and non-citizens.

The most valuable contribution of this study was the discovery that there is no difference between citizens and non-citizens among their preferences on the house features that they prefer, and the most the features they can benefit from while living in the house. For example, flexibility to quickly add or remove spaces, and flexibility to quickly change the elements of the houses. The study also revealed no significant difference between citizens and non-citizens among the house type; both of the groups prefer the *non-constructed house*. The study yielded the following findings:

- The highest motivation and reason for the citizens and non-citizens to buy a constructed house is its layout and appearance, with 48.5% and 39.8% respectively.
- The highest motivation and reason for the citizens and non-citizens to build a non-constructed house is the fixable design that meet their needs (i.e. number of rooms), with 57.6% and 36.9% respectively.
- More than 50% of the participants (citizens and non-citizens) are interested in all of the suggested features.
- The feature that most citizens were interested in is the flexibility to add or remove spaces, with 93.9%, while the most feature that the non-citizens interested in is the flexibility to change the house elements, with 89.6%.

The results revealed a good information to improve the construction method in the housing industry regardless of the location of the house and the surrounding services, which is not included in this study. The findings of this study could be used to guide the industry's leaders and governments to apply a new and different construction method for the houses that contribute to meet the Vision 2030. Also, it could help researchers in the construction industry, or in other industries, to study the end users preferences and perceptions, in other locations, among the houses using similar questionnaire.

**Acknowledgment**
This work was supported in part by a grant from UAE University.

## 6. References

[1] Zavei SJ, and Jusan MM. "Exploring housing attributes selection based on Maslow's hierarchy of needs". *Procedia-Social and Behavioral Sciences.* 2012 Jan 1; 42:311-9.

[2] Selim GM, and Gamal M. "Neighborhoods & residential communities: The case of Al-Ain-UAE". *InProceedings of Housing Symposium* 2007 (**Vol. 3**, pp. 343-361).

[3] Walvekar, S. W. S. "A Study on Consumer Perception towards Affordable Housing Loans". *Asian Journal of Managerial Science*, 2018 7(**2**), 41-46.

[4] Balkir I, Ismail FH, and Al-Segini S. "Determinants of Fund Allocations Out of Housing Programs in the UAE". *Journal of* Economic and Administrative Sciences. 2006 Jun 1.

[5] Solak AO, and Kabadayi B. "An Econometric Analysis of Housing Demand in Turkey". *Advances in Management and Applied Economics.* 2016 May 1;6(**3**):47.

[6] Bekmez S, and Özpolat A. "A Comperative Regional Analysis for Housing Demand in Turkey". *Journal of Business and Economics.* 2014;5(**10**):1854-66.

[7] Hashim AE, Samikon SA, Nasir NM, and Ismail N. "Assessing factors influencing performance of Malaysian low-cost public housing in sustainable environment". *Procedia-Social and Behavioral Sciences.* 2012 Jan **1;50:920-7**.

[8] Mohit MA, Ibrahim M, and Rashid YR. "Assessment of residential satisfaction in newly designed public low-cost housing in Kuala Lumpur, Malaysia". *Habitat international.* 2010 Jan 1;34(**1**):18-27.

[9] Savasdisara T, Tips WE, and Suwannodom S. "Residential satisfaction in private estates in Bangkok: A comparison of low-cost housing estates and determinant factors*". *Habitat International.* 1989 Jan 1;13(**1**):65-73.

[10] Sulaiman H, and Yahaya N. "Housing provision and satisfaction of low-income households in Kuala Lumpur". *Habitat international*. 1987 Jan 1;11(**4**):27-38.

[11] Salleh AG. "Neighbourhood factors in private low-cost housing in Malaysia". *Habitat International*. 2008 Dec 1;32(**4**):485-93.

[12] Burgan BA, Sansom MR. Sustainable steel construction. *Journal of constructional steel research*. 2006 Nov 1;62(**11**):1178-83.

IOP Conf. Series: Materials Science and Engineering 1044 (2021) 012015 doi:10.1088/1757-899X/1044/1/012015

# Integration of Renewable Energy Technologies in University Campuses: a Case Study

**Farnush Nazipov[1], Serik Tokbolat[2,*], Diana Khussainova[1], Mohamad Mustafa[3] and Aziz Almahadin[4]**

[1]Department of Civil and Environmental Engineering, Nazarbayev University, Nur-Sultan, Kazakhstan
[2]Department of Construction Management, School of Architecture Design and the Built Environment, Nottingham Trent University, Nottingham, United Kingdom
[3]Department of Buildings, Energy and Material Technology, UiT The Arctic University of Norway, Narvik, Norway
[4]Aviation Engineering Department, Higher Colleges of Technology, Dubai, UAE

E-mail: stokbolat@gmail.com

**Abstract.** Renewable energy technologies are becoming widely integrated into energy generation systems around the world. While the majority of such technologies are integrated at a large scale in the form of wind and solar farms, there is a need for their incorporation into the built environment. A university campus is an ideal place for testing and understanding how renewable energy technologies can be integrated at a building site scale. Nazarbayev University in Nur-Sultan, Kazakhstan, is a research intense institution which is set to lead the sustainability-related research in the region. The campus is populated with a number of technologies such as a passive house, wind turbines, and photovoltaic solar panels of different capacities and configuration. The review of the technologies and assessment of their operation allowed the study to understand the nature of pioneering technologies that can be integrated in the climate conditions of the locality and how such solutions can help to increase the awareness and popularize renewable energy through education and research.

## 1. Introduction

Energy is a fundamental resource to sustain life on Earth. It supplies cities with light, empowers vehicles, cools or heats the buildings, and has many other applications. Over the centuries, the global population has been relying on energy to advance technological development which simultaneously became a reason for environmental degradation [1-3]. Global warming caused by excessive discharge of greenhouse gas emissions into the atmosphere and depletion of energy resources pose a great threat to the future of the planet. In recent decades, scientists and governments started to search for alternative clean energy sources to meet the growing energy demand and reduce the impacts on global climate [4-7]. At present, the renewable energy output is only a small fraction of all produced energy globally, and it is particularly small in the case of developing countries such as Kazakhstan [8-11]. Therefore, any attempt to integrate renewable energy sources both at a large and a small scale should be based on detailed studies, analysis of lessons learned and investigation of further multiplication possibilities.

A university campus is a suitable environment where sustainability measures and renewable energy technologies can be implemented and studied. This is particularly true due to the availability of researchers, students, and technicians who have time and resources to conduct time-consuming investigations on the efficiency and practicability of renewable energy technologies [2-3]. University campuses with their technological parks and polygons tend to be used as demonstration sites both for students and visitors from outside the campus ranging from government representatives to business owners. This provides an opportunity to raise awareness among various stakeholders who can potentially be engaged in similar projects outside the campus [12-13]. Physical viewing of wind turbines, solar panels and solar water heaters by various stakeholders, and where students are actually engaged with technologies, tend to be more impactful and educational compared to mere theoretical knowledge. Thus, installing renewable energy technologies in university campuses allows increasing the awareness among stakeholders and allows testing their integration in the city scale since a university campus can be considered as a model of a typical urban area. Therefore, a university campus is an ideal place to encourage the promotion of energy sustainability and the integration of renewable energy technologies [2-3]. Many educational institutions around the globe have already integrated renewable energy technologies on their campuses. For example, in the University of Minnesota Morris 70% of daily electricity usage on campus is coming from renewable sources. They have two wind turbine towers with a total generation capacity of more than 20 million kWh of electricity per annum, 20 kW Solar PV Microgrid that powers one of the residence halls and biomass gasification facilities that extracts fossil fuels from local corn and cobs biomass [14]. Colorado State University installed 6.7MW capacity solar PV panels during 2009-2015. Arizona State University installed 24.1 MW on-site and 28.8 MW off-site solar PV panels. University of California campuses are home to 36 MW capacity solar PV panels with more than 52 million kWh electricity [15].

Chinese university Shenyang has taken serious steps in becoming a sustainable university by means of installation of a ground heat pump, recycling of greywater, implementing effective waste management and sustainability-oriented education, and launched many on-going projects on environmental protection [16]. The University of Kansas took green initiatives and installed an energy system that has 36 solar PV panels which produce 7.4 kWh energy and a wind turbine that generates 400 kW per month [17]. The Savona university campus located in Italy meets campus thermal energy demands by installing 28kW cogeneration gas turbines, two 400kW boilers which are fed by natural gas, biomass gasifier connected with an internal combustion engine, 100kW gas turbine microgrid, 250kW diesel engine charged with turbo. Some other initiatives have smart polygeneration microgrid (SPM) that includes wind turbines, PV solar panels, electrical battery storage, two recharge stations for electric vehicles, etc. [18]. University of Cape Town located in South Africa has created a Green Campus Plan and undertook measures such as abatement of carbon emission, conservation of water, and waste recycling system [19]. Kalinga Institute of Social Sciences (KISS) located in India undertook massive sustainable infrastructure initiatives to diminish the carbon footprint of more than 30000 students and faculty members. KISS installed 870KWp Solar PV system on campus that meets 100 percent energy demand and is considered as the largest roof installation amid Indian universities, rooftop solar water heating installation, biogas plant with 1000 kg capacity per day, rainwater harvesting system, etc. [20].

## 2. Review of renewable energy technologies at Nazarbayev University campus

### *2.1. Passive house*

One of the main components of the University's renewable energy polygon is a traditionally shaped "Yurt" building which is built according to the standards of a passive house (Figure 1). It is an energy-efficient autonomous building with a unique round shape, designed in the form of a traditional nomadic house, that has integrated state-of-the-art energy-efficient and renewable energy technologies. The building is powered by solar, wind, and geothermal energy with automated control of the system.

**Figure 1.** Energy efficient "Yurt" building

Passive house design in the case of the "Yurt" building emphasizes harnessing energy from renewable sources. This is also done by optimizing the building design in a way that maximum solar radiation can be used to heat the building, wind to cool the buildings, vegetation to create shadow, and the temperature of the ground to heat and cool the building via a geothermal pump. It is reported that passive design measures allowed the "Yurt" building to perform 3.5 times energy-wise compared to buildings designed and built according to conventional standards [21]. One of the key building design elements that are deemed to create a passive house effect is the shape of the "Yurt" building's roof which is made in the form of a dome. It is reported to be efficient against winds and snow loads, minimizes heat losses, and requires fewer materials and labor to build. It was reported that yurt-shaped buildings use 23-27 percent less energy during exploitation [21]. The main design specifications are presented in Table 1.

**Table 1.** Specifications of the "Yurt" building

| | |
|---|---|
| **Total area (sq.m)** | 115 |
| **Number of floors** | 2 |
| **Number of rooms** | 5 |
| **Building diameter (m)** | 10 |
| **Building height (m)** | 6 |
| **Foundation** | Pile type |
| **Utilities system** | Power, water, drainage |
| **Insulation** | Polyurethane-based |
| **Framework of the dome** | Softwood lumber lined with sheets of OSB |

### *2.2. Photovoltaic (PV) solar panels*

PV solar panels (ZDNY-250C60) is a system with the total number of 15 PV connected solar panels (Figure 2). Each module has dimensions of 1650 x 992 x 45 mm and consists of 60 cells connected in series. One mono-crystalline cell's dimension is 156 x156mm. The weight of the PV solar module is 22.5 kg. The maximum power capacity of the module is 250 Watt and the whole system has a power capacity of 2.5 kW. The efficiency of cells in this model is 17.9%. The cells can operate in -40 to 85 degrees Celsius temperature range and have 47 degrees Celsius nominal operating temperature. Typically, on sunny days, the system produces 7.5-10.0 kW of energy per hour [22]. This energy goes through optimized three phase inverter (Suntellite ZDNY-TL), gets collected in the rechargeable batteries and after that used to supply the autonomous energy building (Yurt) [23].

**Figure 2.** PV solar panels connected to the "Yurt" building

### *2.3. Wind turbine*

The wind turbine made of reinforced fiberglass (Figure 3). It has a maximum power output of 3 kW and has a rated power of 2.5 kW at a wind speed of 11 m/s. The maximum allowable wind speed for the system is 40 m/s with auto furl and dump load for over-speed protection. The operational temperature range is between -30°C and 60°C [24]. The machinery can generate 324 kWh per month at an average wind speed of 5.4 m/s. For the average wind speed of 4.45 m/s in Nur-Sultan, the turbine can produce 267kWh per month. This power is directly supplied to meet the energy demand of the "Yurt" building. The turbine consists of 3 blades with a diameter of 3.8m and it performs 450 rpm at 2.5kW power [24].

**Figure 3.** Wind turbines

## 3. Discussion and Conclusions

As described in the sections above, Nazarbayev University campus has a polygon with integrated renewable energy technologies. Although there is a wide number of technologies in place, this study considered only those which are connected to the "Yurt" building, an energy-efficient house that was designed and built according to passive house principles. The "Yurt" building is also designed with consideration of local traditional design elements pertinent to nomadic style. This design, including the rounded shape and a dome-shaped roof, proved to increase the energy efficiency by exploiting the wind and solar energy effectively. The "Yurt" building is connected to the network of 15 solar PV panels that produce 7.5-10.0 kW of energy per hour and a wind turbine that can produce 267kWh per month. The "Yurt" building itself reported being 3.5 times more energy efficient compared to standard houses built according to conventional construction norms. The energy produced by the integration of renewable energy and saved by the passive design contributes to the overall sustainability of the university campus. The passive house is a unique concept for Kazakhstan and can be considered as a breakthrough project which needs to be studied more carefully and replicated at a wider scale. The same can be stated regarding the wind turbine and solar PV panels. Existing wind turbines and solar panels supply energy to batteries and not to actual homes; so, it can be stated that the case study of "Yurt" building with the integrated renewable energy technologies can be considered as a pioneering model of a smart grid. Careful examination and long-term assessment of such a grid should be carried out and replicated at a wider scale.

The future study should discuss all the renewable energy technologies installed at Nazarbayev University campus such as Concentrated Solar Power (CSP) plant - Sun Power (25 kW), PV solar panels (15 kW) integrated to the Technopark building, tiltable solar tracker (10 kW), geothermal pump (3 kW), and solar collectors (32 sq.m.). One of the limitations of the study can be related to the fact that the operational data on monthly energy generation over the past periods was not available for the analysis due to commercial privacy. Future research should also consider the cost-benefit analysis to understand the payback period of building a passive house and integrating various renewable energy technologies.

**Acknowledgement**
This research was supported by the Project: Academic Cooperation in Post-graduate Engineering Education (ACE) - CPEA-ST-2019/10029 funded by the Norwegian Agency for International Cooperation and Quality Enhancement in Higher Education (Diku) under Eurasia Programme 2019. We would also like to thank Nazarbayev University Research and Innovation Systems (NURIS) for providing valuable information on the renewable energy technologies at Nazarbayev University campus.

## 4 References

[1] Ahmad, S., Nadeem, A., Akhanova, G., Houghton, T., & Muhammad-Sukki, F. (2017). Multi-criteria evaluation of renewable and nuclear resources for electricity generation in Kazakhstan. Energy, 141, 1880-1891. doi.org/10.1016/j.energy.2017.11.102

[2] Kalkan, N., Bercin, K., Cangul, O., Morales, M. G., Saleem, M. M. K. M., Marji, I., ... & Tsigkogianni, E. (2011). A renewable energy solution for Highfield Campus of University of Southampton. *Renewable and sustainable energy reviews*, *15*(6), 2940-2959. doi.org/10.1016/j.rser.2011.02.040

[3] Islam, M. S., Hoque, N., & Beg, M. R. A. (2017). ICMEAS 2017 Feasibility Study of Implementing Renewable Energy on an institutional Campus in Bangladesh–A case study. *Military Institute of Science and Technology: Dhaka, Bangladesh.*

[4] Demirbas, A. (2009). Global renewable energy projections. Energy Sources, Part B, 4(2), 212-224. doi.org/10.1080/15567240701620499

[5] Enerdata. (2013). Global energy statistical yearbook. www.enerdata.net

[6] Ellabban, O., Abu-Rub, H., & Blaabjerg, F. (2014). Renewable energy resources: Current status, future prospects and their enabling technology. Renewable and Sustainable Energy Reviews, 39, 748-764. doi.org/10.1016/j.rser.2014.07.113

[7] Panwar, N. L., Kaushik, S. C., & Kothari, S. (2011). Role of renewable energy sources in environmental protection: A review. Renewable and sustainable energy reviews, 15(3), 1513-1524. doi.org/10.1016/j.rser.2010.11.037

[8] Karatayev, M., & Clarke, M. L. (2014). Current energy resources in Kazakhstan and the future potential of renewables: A review. Energy Procedia, 59, 97-104. doi.org/10.1016/j.egypro.2014.10.354

[9] Assembayeva, M., Zhakiyev, N., & Akhmetbekov, Y. (2017). Impact of storage technologies on renewable energy integration in Kazakhstan. Materials Today: Proceedings, 4(3), 4512-4523. doi.org/10.1016/j.matpr.2017.04.024

[10] Karatayev, M., Hall, S., Kalyuzhnova, Y., & Clarke, M. L. (2016). Renewable energy technology uptake in Kazakhstan: Policy drivers and barriers in a transitional economy. Renewable and Sustainable Energy Reviews, 66, 120-136. doi.org/10.1016/j.rser.2016.07.057

[11] Karatayev, M., & Clarke, M. L. (2016). A review of current energy systems and green energy potential in Kazakhstan. Renewable and Sustainable Energy Reviews, 55, 491-504. doi.org/10.1016/j.rser.2015.10.078

[12] Akindeji, K. T., Tiako, R., & Davidson, I. E. (2019). Use of Renewable Energy Sources in University Campus Microgrid–A Review. In *2019 International Conference on the Domestic Use of Energy (DUE)*:76-83. IEEE.
[13] Sonetti, G., Lombardi, P., & Chelleri, L. (2016). True green and sustainable university campuses? Toward a clusters approach. *Sustainability*, *8*(1), 83. doi.org/10.3390/su8010083
[14] Renewable Energy. (2020). *University of Minnesota, Morris*. https://morris.umn.edu/sustainability/renewable-energy.
[15] Elgqvist, E. M., & Van Geet, O. D. (2017). *Campus Energy Approach, REopt Overview, and Solar for Universities* (No. NREL/PR-7A40-70252). National Renewable Energy Lab.(NREL), Golden, CO (United States).
[16] Geng, Y., Liu, K., Xue, B., & Fujita, T. (2013). Creating a "green university" in China: a case of Shenyang University. *Journal of Cleaner Production*, *61*, 13-19. doi.org/10.1016/j.jclepro.2012.07.013
[17] Murray, Ryan et al. (2014). "A Solar Energy Proposal for the University of Kansas". Capstone Project, University of Kansas.
[18] Bracco, S., Delfino, F., Pampararo, F., Robba, M., & Rossi, M. (2013). The University of Genoa smart polygeneration microgrid test-bed facility: The overall system, the technologies and the research challenges. *Renewable and sustainable energy reviews*, *18*, 442-459. doi.org/10.1016/j.rser.2012.10.009
[19] Hall, M., & Murray, N. (2008). Green Campus Policy Framework for the University of Cape Town. *University of Cape Town, Cape Town*.
[20] *ANNUAL REPORT 2018-19*. (2020). https://kiss.ac.in/pdf/FINAL-2019-Annual-Report.pdf.
[21] Tyo, V., & Yessengabulov, S. (2015). A Comparative Case Study of the Impact of Square and Yurt-Shape Buildings on Energy Efficiency. International Journal of Environmental and Ecological Engineering, 9(10), 1214-1217. doi.org/10.5281/zenodo.1108881
[22] Electrotecenergy.se. http://electrotecenergy.se/wp-content/uploads/2014/11/Solcellspanel-Suntellite-mono.pdf.
[23] Nusolar.com.my. http://www.nusolar.com.my/pdfs/ZDNY-TL%2010000-17000.pdf.
[24] Yueniao.com. http://www.yueniao.com/HY-5KW.html.

# AUTHOR INDEX